Wifi Technology:
Advances and Applications

Wifi Technology: Advances and Applications

Editor: Jacob Davis

NY RESEARCH
P R E S S

New York

Published by NY Research Press
118-35 Queens Blvd., Suite 400,
Forest Hills, NY 11375, USA
www.nyresearchpress.com

Wifi Technology: Advances and Applications
Edited by Jacob Davis

International Standard Book Number: 978-1-63238-598-7 (Hardback)

Cataloging-in-Publication Data

Wifi technology : advances and applications / edited by Jacob Davis.
     p. cm.
Includes bibliographical references and index.
ISBN 978-1-63238-598-7
1. Wireless LANs. 2. Wireless communication systems. I. Davis, Jacob.
TK5105.78 .W54 2018
621.382 1--dc23

Contents

Permissions

Index

Preface

Today, life in many parts of the world cannot be imagined without Wi-Fi technology. It has made the experience of Internet very unique and convenient. Wi-Fi technology is wireless local area networking. It enables devices like smartphones, laptops, digital cameras, computers, modern printers, tablets, etc. to access Internet through wireless access point. This book presents the complex subject of Wi-Fi technology in the most comprehensible and easy to understand language. It is compiled in such a manner, that it will provide in-depth knowledge about the theory and practice of the subject. Coherent flow of topics, student-friendly language and extensive use of examples make this textbook an invaluable source of knowledge.

A short introduction to every chapter is written below to provide an overview of the content of the book:

Chapter 1 - WiFi is a wireless technology which uses IEEE 802.11 standards. It can be used in computers, laptops, mobile phones and smart TVs. This chapter is an overview of the subject matter incorporating all the major aspects of WiFi; **Chapter 2 -** Wireless networks are computer networks which do not require cables. They are mainly administered by using radio communication. Wireless mesh network, free-space optical communication, municipal wireless network, wireless network interface controller, fixed wireless and wireless security are important topics related to the subject matter; **Chapter 3 -** A wireless LAN is used in connecting two or more devices to a wireless network. The types of wireless LANs are infrastructure, peer-to-peer, wireless distribution system and bridge. The chapter strategically encompasses and incorporates the major components and key concepts of wireless LAN, providing a complete understanding; **Chapter 4 -** Wireless sensor networks are mainly built of nodes. The applications can be categorized into area monitoring, environmental sensing, health care monitoring and industrial monitoring. Sensor node, mobile wireless sensor network and optical wireless communications have been explained in the chapter. WiFi technology is best understood in confluence with the major topics listed in the following chapter; **Chapter 5 -** Various devices that enable wireless networking include wireless router, MiFi, Mobile broadband modem, etc. A wireless router sends data packets using WLAN as well as provides access to a wired network. Tools and techniques are an important component of any field of study. The following chapter elucidates the various tools and techniques that are related to WiFi.

I extend my sincere thanks to the publisher for considering me worthy of this task. Finally, I thank my family for being a source of support and help.

Editor

An Introduction to WiFi

WiFi is a wireless technology which uses IEEE 802.11 standards. It can be used in computers, laptops, mobile phones and smart TVs. This chapter is an overview of the subject matter incorporating all the major aspects of WiFi.

Wi-Fi

Wi-Fi or WiFi is a technology for wireless local area networking with devices based on the IEEE 802.11 standards. *Wi-Fi* is a trademark of the Wi-Fi Alliance, which restricts the use of the term *Wi-Fi Certified* to products that successfully complete interoperability certification testing.

Devices that can use Wi-Fi technology include personal computers, video-game consoles, smartphones, digital cameras, tablet computers, smart TVs, digital audio players and modern printers. Wi-Fi compatible devices can connect to the Internet via a WLAN and a wireless access point. Such an access point (or hotspot) has a range of about 20 meters (66 feet) indoors and a greater range outdoors. Hotspot coverage can be as small as a single room with walls that block radio waves, or as large as many square kilometres achieved by using multiple overlapping access points.

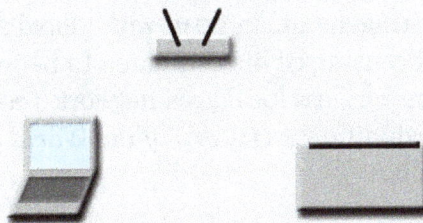

Depiction of a device sending information wirelessly to another device, both connected to the local network, in order to print a document

Wi-Fi most commonly uses the 2.4 gigahertz (12 cm) UHF and 5 gigahertz (6 cm) SHF ISM radio bands. Having no physical connections, it is more vulnerable to attack than wired connections, such as Ethernet.

History

In 1971, ALOHAnet connected the Hawaiian Islands with a UHF wireless packet network. ALOHAnet and the ALOHA protocol were early forerunners to Ethernet, and later the IEEE 802.11 protocols, respectively.

A 1985 ruling by the U.S. Federal Communications Commission released the ISM band for unlicensed use. These frequency bands are the same ones used by equipment such as microwave ovens and are subject to interference.

In 1991, NCR Corporation with AT&T Corporation invented the precursor to 802.11, intended for use in cashier systems, under the name WaveLAN.

The Australian radio-astronomer Dr John O'Sullivan with his colleagues Terence Percival, Graham Daniels, Diet Ostry, John Deane developed a key patent used in Wi-Fi as a by-product of a Commonwealth Scientific and Industrial Research Organisation (CSIRO) research project, "a failed experiment to detect exploding mini black holes the size of an atomic particle". Dr O'Sullivan and his colleagues are credited with inventing Wi-Fi. In 1992 and 1996, CSIRO obtained patents for a method later used in Wi-Fi to "unsmear" the signal.

The first version of the 802.11 protocol was released in 1997, and provided up to 2 Mbit/s link speeds. This was updated in 1999 with 802.11b to permit 11 Mbit/s link speeds, and this proved to be popular.

In 1999, the Wi-Fi Alliance formed as a trade association to hold the Wi-Fi trademark under which most products are sold.

Wi-Fi uses a large number of patents held by many different organizations. In April 2009, 14 technology companies agreed to pay CSIRO $1 billion for infringements on CSIRO patents. This led to Australia labeling Wi-Fi as an Australian invention, though this has been the subject of some controversy. CSIRO won a further $220 million settlement for Wi-Fi patent-infringements in 2012 with global firms in the United States required to pay the CSIRO licensing rights estimated to be worth an additional $1 billion in royalties. In 2016, the wireless local area network Test Bed was chosen as Australia's contribution to the exhibition *A History of the World in 100 Objects* held in the National Museum of Australia.

Etymology

The name *Wi-Fi*, commercially used at least as early as August 1999, was coined by the brand-consulting firm Interbrand. The Wi-Fi Alliance had hired Interbrand to create a name that was "a little catchier than 'IEEE 802.11b Direct Sequence.'" Phil Belanger, a founding member of the Wi-Fi Alliance who presided over the selection of the name "Wi-Fi," has stated that Interbrand invented *Wi-Fi* as a pun upon the word *hi-fi*.

Interbrand also created the Wi-Fi logo. The yin-yang Wi-Fi logo indicates the certification of a product for interoperability.

The Wi-Fi Alliance used the nonsense advertising slogan "The Standard for Wireless Fidelity" for a short time after the brand name was created. The name was however never officially "Wireless Fidelity". Nevertheless, the Wi-Fi Alliance was also called the "Wireless Fidelity Alliance Inc" in some publications and the IEEE's own website has stated "WiFi is a short name for Wireless Fidelity".

Non-Wi-Fi technologies intended for fixed points, such as Motorola Canopy, are usually described as fixed wireless. Alternative wireless technologies include mobile phone standards, such as 2G, 3G, 4G, and LTE.

The name is sometimes written as WiFi, Wifi, or wifi, but these are not approved by the Wi-Fi Alliance.

Wi-Fi ad-hoc Mode

Wi-Fi nodes operating in ad-hoc mode refers to devices talking directly to each other without the need to first talk to an access point (also known as base station). Ad-hoc mode was first invented and realized by Chai K. Toh in his 1996 invention of Wi-Fi ad-hoc routing, implemented on Lucent WaveLAN 802.11a wireless on IBM ThinkPads over a size nodes scenario spanning a region of over a mile. The success was recorded in Mobile Computing magazine (1999) and later published formally in IEEE Transactions on Wireless Communications, 2002 and ACM SIGMETRICS Performance Evaluation Review, 2001.

Wi-Fi Certification

The IEEE does not test equipment for compliance with their standards. The non-profit Wi-Fi Alliance was formed in 1999 to fill this void — to establish and enforce standards for interoperability and backward compatibility, and to promote wireless local-area-network technology. As of 2010, the Wi-Fi Alliance consisted of more than 375 companies from around the world. The Wi-Fi Alliance enforces the use of the Wi-Fi brand to technologies based on the IEEE 802.11 standards from the IEEE. This includes wireless local area network (WLAN) connections, device to device connectivity (such as Wi-Fi Peer to Peer aka Wi-Fi Direct), Personal area network (PAN), local area network (LAN) and even some limited wide area network (WAN) connections. Manufacturers with membership in the Wi-Fi Alliance, whose products pass the certification process, gain the right to mark those products with the Wi-Fi logo.

Specifically, the certification process requires conformance to the IEEE 802.11 radio standards, the WPA and WPA2 security standards, and the EAP authentication standard. Certification may optionally include tests of IEEE 802.11 draft standards, interaction with cellular-phone technology in converged devices, and features relating to security set-up, multimedia, and power-saving.

Not every Wi-Fi device is submitted for certification. The lack of Wi-Fi certification does not necessarily imply that a device is incompatible with other Wi-Fi devices. The Wi-Fi Alliance may or may not sanction derivative terms, such as Super Wi-Fi, coined by the US Federal Communications Commission (FCC) to describe proposed networking in the UHF TV band in the US.

IEEE 802.11 Standard

This Netgear Wi-Fi router contains dual bands for transmitting the 802.11 standard across the 2.4 and 5 GHz spectrums.

The IEEE 802.11 standard is a set of media access control (MAC) and physical layer (PHY) specifications for implementing wireless local area network (WLAN) computer communication in the 2.4, 3.6, 5, and 60 GHz frequency bands. They are created and maintained by the IEEE LAN/MAN Standards Committee (IEEE 802). The base version of the standard was released in 1997, and has had subsequent amendments. The standard and amendments provide the basis for wireless network products using the Wi-Fi brand. While each amendment is officially revoked when it is incorporated in the latest version of the standard, the corporate world tends to market to the revisions because they concisely denote capabilities of their products. As a result, in the market place, each revision tends to become its own standard.

Uses

A Japanese sticker indicating to the public that a location is within range of a Wi-Fi network. A dot with curved lines radiating from it is a common symbol for Wi-Fi, representing a point transmitting a signal.

To connect to a Wi-Fi LAN, a computer has to be equipped with a wireless network interface controller. The combination of computer and interface controller is called a *station*. For all stations that share a single radio frequency communication channel, transmissions on this channel are received by all stations within range. The transmission is not guaranteed to be delivered and is therefore a best-effort delivery mechanism. A carrier wave is used to transmit the data. The data is organised in packets on an Ethernet link, referred to as "Ethernet frames".

Internet Access

Wi-Fi technology may be used to provide Internet access to devices that are within the range of a wireless network that is connected to the Internet. The coverage of one or more interconnected access points (*hotspots*) can extend from an area as small as a few rooms to as large as many square kilometres. Coverage in the larger area may require a group of access points with overlapping coverage. For example, public outdoor Wi-Fi technology has been used successfully in wireless mesh networks in London, UK. An international example is Fon.

Wi-Fi provides service in private homes, businesses, as well as in public spaces at Wi-Fi hotspots set up either free-of-charge or commercially, often using a captive portal webpage for access. Organizations and businesses, such as airports, hotels, and restaurants, often provide free-use hotspots to attract customers. Enthusiasts or authorities who wish to provide services or even to promote business in selected areas sometimes provide free Wi-Fi access.

Routers that incorporate a digital subscriber line modem or a cable modem and a Wi-Fi access point, often set up in homes and other buildings, provide Internet access and internetworking to all devices connected to them, wirelessly or via cable.

Similarly, battery-powered routers may include a cellular Internet radiomodem and Wi-Fi access point. When subscribed to a cellular data carrier, they allow nearby Wi-Fi stations to access the Internet over 2G, 3G, or 4G networks using the tethering technique. Many smartphones have a built-in capability of this sort, including those based on Android, BlackBerry, Bada, iOS (iPhone), Windows Phone and Symbian, though carriers often disable the feature, or charge a separate fee to enable it, especially for customers with unlimited data plans. "Internet packs" provide standalone facilities of this type as well, without use of a smartphone; examples include the MiFi- and Wi-Bro-branded devices. Some laptops that have a cellular modem card can also act as mobile Internet Wi-Fi access points.

Wi-Fi also connects places that normally don't have network access, such as kitchens and garden sheds.

Google is intending to use the technology to allow rural areas to enjoy connectivity by utilizing a broad mix of projection and routing services. Google also intends to bring connectivity to Africa and some Asian lands by launching blimps that will allow for internet connection with Wi-Fi technology.

City-wide Wi-Fi

An outdoor Wi-Fi access point

In the early 2000s, many cities around the world announced plans to construct city-wide Wi-Fi networks. There are many successful examples; in 2004, Mysore became India's first Wi-Fi-enabled city. A company called WiFiyNet has set up hotspots in Mysore, covering the complete city and a few nearby villages.

In 2005, St. Cloud, Florida and Sunnyvale, California, became the first cities in the United States to offer citywide free Wi-Fi (from MetroFi). Minneapolis has generated $1.2 million in profit annually for its provider.

In May 2010, London, UK, Mayor Boris Johnson pledged to have London-wide Wi-Fi by 2012. Several boroughs including Westminster and Islington already had extensive outdoor Wi-Fi coverage at that point.

Officials in South Korea's capital are moving to provide free Internet access at more than 10,000 locations around the city, including outdoor public spaces, major streets and densely populated residential areas. Seoul will grant leases to KT, LG Telecom and SK Telecom. The companies will invest $44 million in the project, which was to be completed in 2015.

Campus-wide Wi-Fi

Many traditional university campuses in the developed world provide at least partial Wi-Fi coverage. Carnegie Mellon University built the first campus-wide wireless Internet network, called Wireless Andrew, at its Pittsburgh campus in 1993 before Wi-Fi branding originated. By February 1997 the CMU Wi-Fi zone was fully operational. Many universities collaborate in providing Wi-Fi access to students and staff through the Eduroam international authentication infrastructure.

Wi-Fi ad hoc Versus Wi-Fi direct

Wi-Fi also allows communications directly from one computer to another without an access point intermediary. This is called *ad hoc* Wi-Fi transmission. This wireless ad hoc network mode has proven popular with multiplayer handheld game consoles, such as the Nintendo DS, PlayStation Portable, digital cameras, and other consumer electronics devices. Some devices can also share their Internet connection using ad hoc, becoming hotspots or "virtual routers".

Similarly, the Wi-Fi Alliance promotes the specification Wi-Fi Direct for file transfers and media sharing through a new discovery- and security-methodology. Wi-Fi Direct launched in October 2010.

Another mode of direct communication over Wi-Fi is Tunneled Direct Link Setup (TDLS), which enables two devices on the same Wi-Fi network to communicate directly, instead of via the access point.

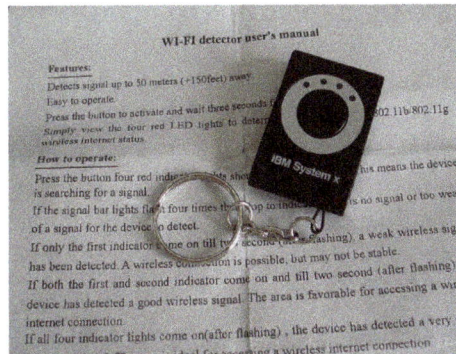

A keychain-size Wi-Fi detector

Wi-Fi Radio Spectrum

802.11b and 802.11g use the 2.4 GHz ISM band, operating in the United States under Part 15 Rules and Regulations. Because of this choice of frequency band, 802.11b and g equipment may occasionally suffer interference from microwave ovens, cordless telephones, and Bluetooth devices.

Spectrum assignments and operational limitations are not consistent worldwide: Australia and Europe allow for an additional two channels (12, 13) beyond the 11 permitted in the United States for the 2.4 GHz band, while Japan has three more (12–14). In the US and other countries, 802.11a and 802.11g devices may be operated without a license, as allowed in Part 15 of the FCC Rules and Regulations.

A Wi-Fi signal occupies five channels in the 2.4 GHz band. Any two channel numbers that differ by five or more, such as 2 and 7, do not overlap. The oft-repeated adage that channels 1, 6, and 11 are the *only* non-overlapping channels is, therefore, not accurate. Channels 1, 6, and 11 are the only *group of three* non-overlapping channels in North

America and the United Kingdom. In Europe and Japan using Channels 1, 5, 9, and 13 for 802.11g and 802.11n is recommended.

802.11a uses the 5 GHz U-NII band, which, for much of the world, offers at least 23 non-overlapping channels rather than the 2.4 GHz ISM frequency band, where adjacent channels overlap.

Interference

Wi-Fi connections can be disrupted or the Internet speed lowered by having other devices in the same area. Many 2.4 GHz 802.11b and 802.11g access-points default to the same channel on initial startup, contributing to congestion on certain channels. Wi-Fi pollution, or an excessive number of access points in the area, especially on the neighboring channel, can prevent access and interfere with other devices' use of other access points, caused by overlapping channels in the 802.11g/b spectrum, as well as with decreased signal-to-noise ratio (SNR) between access points. This can become a problem in high-density areas, such as large apartment complexes or office buildings with many Wi-Fi access points.

Additionally, other devices use the 2.4 GHz band: microwave ovens, ISM band devices, security cameras, ZigBee devices, Bluetooth devices, video senders, cordless phones, baby monitors, and, in some countries, amateur radio, all of which can cause significant additional interference. It is also an issue when municipalities or other large entities (such as universities) seek to provide large area coverage.

Service Set Identifier (SSID)

In addition to running on different channels, multiple Wi-Fi networks can share channels.

A service set is the set of all the devices associated with a particular Wi-Fi network. The service set can be local, independent, extended or mesh.

Each service set has an associated identifier, the 32-byte Service Set Identifier (SSID), which identifies the particular network. The SSID is configured within the devices that are considered part of the network, and it is transmitted in the packets. Receivers ignore wireless packets from networks with a different SSID.

Throughput

As the 802.11 specifications evolved to support higher throughput, the bandwidth requirements also increased to support them. 802.11n uses double the radio spectrum/bandwidth (40 MHz) compared to 802.11a or 802.11g (20 MHz).[76] This means there can be only one 802.11n network on the 2.4 GHz band at a given location, without interference to/from other WLAN traffic. 802.11n can also be set to limit itself to 20 MHz bandwidth to prevent interference in dense community.

Many newer consumer devices support the latest 802.11ac standard, which uses the 5 GHz band exclusively and is capable of multi-station WLAN throughput of at least 1 gigabit per second, and a single station throughput of at least 500 Mbit/s.. In the first quarter of 2016, The Wi-Fi Alliance certifies devices compliant with the 802.11ac standard as "Wi-Fi CERTIFIED ac". This new standard uses several advanced signal processing techniques such as multi-user MIMO and 4X4 Spatial Multiplexing streams, and large channel bandwidth (160 MHz) to achieve the Gigabit throughput. 70% of all access point sales revenue came from 802.11ac devices.

Hardware

Wi-Fi whitelist triggered on an HP laptop

Wi-Fi allows cheaper deployment of local area networks (LANs). Also, spaces where cables cannot be run, such as outdoor areas and historical buildings, can host wireless LANs. However, building walls of certain materials, such as stone with high metal content, can block Wi-Fi signals.

Manufacturers are building wireless network adapters into most laptops. The price of chipsets for Wi-Fi continues to drop, making it an economical networking option included in even more devices.

Different competitive brands of access points and client network-interfaces can inter-operate at a basic level of service. Products designated as "Wi-Fi Certified" by the Wi-Fi Alliance are backward compatible. Unlike mobile phones, any standard Wi-Fi device will work anywhere in the world.

Standard Devices

A wireless access point (WAP) connects a group of wireless devices to an adjacent wired LAN. An access point resembles a network hub, relaying data between connected wireless devices in addition to a (usually) single connected wired device, most often an Ethernet hub or switch, allowing wireless devices to communicate with other wired devices.

OSBRiDGE 3GN – 802.11n Access Point and UMTS/GSM Gateway in one device.

An Atheros draft-N Wi-Fi adapter with built in Bluetooth on a Sony Vaio E series laptop.

An embedded RouterBoard 112 with U.FL-RS-MA pigtail and R52 mini PCI Wi-Fi card widely used by wireless Internet service providers (WISPs) in the Czech Republic.

USB wireless adapter

Wireless adapters allow devices to connect to a wireless network. These adapters connect to devices using various external or internal interconnects such as PCI, miniPCI, USB, ExpressCard, Cardbus and PC Card. As of 2010, most newer laptop computers come equipped with built in internal adapters.

Wireless routers integrate a Wireless Access Point, Ethernet switch, and internal router firmware application that provides IP routing, NAT, and DNS forwarding through an integrated WAN-interface. A wireless router allows wired and wireless Ethernet LAN devices to connect to a (usually) single WAN device such as a cable modem or a DSL modem. A wireless router allows all three devices, mainly the access point and router, to be configured through one central utility. This utility is usually an integrated web server that is accessible to wired and wireless LAN clients and often optionally to WAN clients. This utility may also be an application that is run on a computer, as is the case with as Apple's AirPort, which is managed with the AirPort Utility on macOS and iOS.

Wireless network bridges connect a wired network to a wireless network. A bridge differs from an access point: an access point connects wireless devices to a wired network at the data-link layer. Two wireless bridges may be used to connect two wired networks over a wireless link, useful in situations where a wired connection may be unavailable, such as between two separate homes or for devices which do not have wireless networking capability (but have wired networking capability), such as consumer entertainment devices; alternatively, a wireless bridge can be used to enable a device which supports a wired connection to operate at a wireless networking standard which is faster than supported by the wireless network connectivity feature (external dongle or inbuilt) supported by the device (e.g. enabling Wireless-N speeds (up to the maximum supported speed on the wired Ethernet port on both the bridge and connected devices including the wireless access point) for a device which only supports Wireless-G). A dual-band wireless bridge can also be used to enable 5 GHz wireless network operation on a device which only supports 2.4 GHz wireless networking functionality and has a wired Ethernet port.

Wireless range-extenders or wireless repeaters can extend the range of an existing wireless network. Strategically placed range-extenders can elongate a signal area or allow for the signal area to reach around barriers such as those pertaining in L-shaped corridors. Wireless devices connected through repeaters will suffer from an increased latency for each hop, as well as from a reduction in the maximum data throughput that is available. In addition, the effect of additional users using a network employing wireless range-extenders is to consume the available bandwidth faster than would be the case where but a single user migrates around a network employing extenders. For this reason, wireless range-extenders work best in networks supporting very low traffic throughput requirements, such as for cases where but a single user with a Wi-Fi equipped tablet migrates around the combined extended and non-extended portions of the total connected network. Additionally, a wireless device connected to any of the repeaters in the chain will have a data throughput that is also limited by the "weakest link" existing in the chain between where the connection originates and where the connection ends. Networks employing wireless extenders are also more prone to degradation from interference from neighboring access points that border portions of the extended network and that happen to occupy the same channel as the extended network.

The security standard, Wi-Fi Protected Setup, allows embedded devices with limited graphical user interface to connect to the Internet with ease. Wi-Fi Protected Setup has 2 configurations: The Push Button configuration and the PIN configuration. These embedded devices are also called The Internet of Things and are low-power, battery-operated embedded systems. A number of Wi-Fi manufacturers design chips and modules for embedded Wi-Fi, such as GainSpan.

Embedded Systems

Increasingly in the last few years (particularly as of 2007), embedded Wi-Fi modules have become available that incorporate a real-time operating system and provide a simple means of wirelessly enabling any device which has and communicates via a serial port. This allows the design of simple monitoring devices. An example is a portable ECG device monitoring a patient at home. This Wi-Fi-enabled device can communicate via the Internet.

Embedded serial-to-Wi-Fi module

These Wi-Fi modules are designed by OEMs so that implementers need only minimal Wi-Fi knowledge to provide Wi-Fi connectivity for their products.

In June 2014 Texas Instruments introduced the first ARM Cortex-M4 microcontroller with an onboard dedicated Wi-Fi MCU, the SimpleLink CC3200. It makes embedded systems with Wi-Fi connectivity possible to build as single-chip devices, which reduces their cost and minimum size, making it more practical to build wireless-networked controllers into inexpensive ordinary objects.

Range

The Wi-Fi signal range depends on the frequency band, radio power output, antenna gain and antenna type as well as the modulation technique. Line-of-sight is the thumbnail guide but reflection and refraction can have a significant impact.

An access point compliant with either 802.11b or 802.11g, using the stock antenna might have a range of 100 m (0.062 mi). The same radio with an external semi parabolic antenna (15 dB gain) might have a range over 20 miles.

Higher gain rating (dBi) indicates further deviation (generally toward the horizontal) from a theoretical, perfect isotropic radiator, and therefore the further the antenna can project a usable signal, as compared to a similar output power on a more isotropic antenna. For example, an 8 dBi antenna used with a 100 mW driver will have a similar horizontal range to a 6 dBi antenna being driven at 500 mW. Note that this assumes that radiation in the vertical is lost; this may not be the case in some situations, especially in large buildings or within a waveguide. In the above example, a directional waveguide could cause the low power 6 dBi antenna to project much further in a single direction than the 8 dBi antenna which is not in a waveguide, even if they are both being driven at 100 mW.

IEEE 802.11n, however, can more than double the range. Range also varies with frequency band. Wi-Fi in the 2.4 GHz frequency block has slightly better range than Wi-Fi in the 5 GHz frequency block used by 802.11a (and optionally by 802.11n). On wireless routers with detachable antennas, it is possible to improve range by fitting upgraded antennas which have higher gain in particular directions. Outdoor ranges can be improved to many kilometers through the use of high gain directional antennas at the router and remote device(s). In general, the maximum amount of power that a Wi-Fi device can transmit is limited by local regulations, such as FCC Part 15 in the US. Equivalent isotropically radiated power (EIRP) in the European Union is limited to 20 dBm (100 mW).

To reach requirements for wireless LAN applications, Wi-Fi has fairly high power consumption compared to some other standards. Technologies such as Bluetooth (designed to support wireless personal area network (PAN) applications) provide a much shorter propagation range between 1 and 100 m and so in general have a lower power consumption. Other low-power technologies such as ZigBee have fairly long range, but much lower data rate. The high power consumption of Wi-Fi makes battery life in mobile devices a concern.

Researchers have developed a number of "no new wires" technologies to provide alternatives to Wi-Fi for applications in which Wi-Fi's indoor range is not adequate and where installing new wires (such as CAT-6) is not possible or cost-effective. For example, the ITU-T G.hn standard for high speed local area networks uses existing home wiring (coaxial cables, phone lines and power lines). Although G.hn does not provide some of the advantages of Wi-Fi (such as mobility or outdoor use), it is designed for applications (such as IPTV distribution) where indoor range is more important than mobility.

For the best performance, a number of people only recommend using wireless networking as a supplement to wired networking.

Due to the complex nature of radio propagation at typical Wi-Fi frequencies, particularly the effects of signal reflection off trees and buildings, algorithms can only approximately predict Wi-Fi signal strength for any given area in relation to a transmitter. This effect does not apply equally to long-range Wi-Fi, since longer links typically operate from towers that transmit above the surrounding foliage.

The practical range of Wi-Fi essentially confines mobile use to such applications as inventory-taking machines in warehouses or in retail spaces, barcode-reading devices at checkout stands, or receiving/shipping stations. Mobile use of Wi-Fi over wider ranges is limited, for instance, to uses such as in an automobile moving from one hotspot to another. Other wireless technologies are more suitable for communicating with moving vehicles.

Distance Records

Distance records (using non-standard devices) include 382 km (237 mi) in June 2007, held by Ermanno Pietrosemoli and EsLaRed of Venezuela, transferring about 3 MB of data between the mountain-tops of El Águila and Platillon. The Swedish Space Agency transferred data 420 km (260 mi), using 6 watt amplifiers to reach an overhead strato-spheric balloon.

Multiple Access Points

Increasing the number of Wi-Fi access points provides network redundancy, better range, support for fast roaming and increased overall network-capacity by using more channels or by defining smaller cells. Except for the smallest implementations (such as home or small office networks), Wi-Fi implementations have moved toward "thin" access points, with more of the network intelligence housed in a centralized network appliance, relegating individual access points to the role of "dumb" transceivers. Outdoor applications may use mesh topologies.

When multiple access points are deployed they are often configured with the same SSID and security settings to form an "extended service set". Wi-Fi client devices will typically connect to the access point that can provide the strongest signal within that service set.

Network Security

The main issue with wireless network security is its simplified access to the network compared to traditional wired networks such as Ethernet. With wired networking, one must either gain access to a building (physically connecting into the internal network), or break through an external firewall. To enable Wi-Fi, one merely needs to be within the range of the Wi-Fi network. Most business networks protect sensitive data and systems by attempting to disallow external access. Enabling wireless connectivity reduces security if the network uses inadequate or no encryption.

An attacker who has gained access to a Wi-Fi network router can initiate a DNS spoofing attack against any other user of the network by forging a response before the queried DNS server has a chance to reply.

Securing Methods

A common measure to deter unauthorized users involves hiding the access point's name by disabling the SSID broadcast. While effective against the casual user, it is ineffective as a security method because the SSID is broadcast in the clear in response to a client SSID query. Another method is to only allow computers with known MAC addresses to join the network, but determined eavesdroppers may be able to join the network by spoofing an authorized address.

Wired Equivalent Privacy (WEP) encryption was designed to protect against casual snooping but it is no longer considered secure. Tools such as AirSnort or Aircrack-ng can quickly recover WEP encryption keys. Because of WEP's weakness the Wi-Fi Alliance approved Wi-Fi Protected Access (WPA) which uses TKIP. WPA was specifically designed to work with older equipment usually through a firmware upgrade. Though more secure than WEP, WPA has known vulnerabilities.

The more secure WPA2 using Advanced Encryption Standard was introduced in 2004 and is supported by most new Wi-Fi devices. WPA2 is fully compatible with WPA.

A flaw in a feature added to Wi-Fi in 2007, called Wi-Fi Protected Setup (WPS), allows WPA and WPA2 security to be bypassed and effectively broken in many situations. The only remedy as of late 2011 is to turn off Wi-Fi Protected Setup, which is not always possible.

Virtual Private Networks are often used to secure Wi-Fi.

Data Security Risks

The older wireless encryption-standard, Wired Equivalent Privacy (WEP), has been shown to be easily breakable even when correctly configured. Wi-Fi Protected Access (WPA and WPA2) encryption, which became available in devices in 2003, aimed to solve this problem. Wi-Fi access points typically default to an encryption-free (*open*) mode. Novice users benefit from a zero-configuration device that works out-of-the-box, but this default does not enable any wireless security, providing open wireless access to a LAN. To turn security on requires the user to configure the device, usually via a software graphical user interface (GUI). On unencrypted Wi-Fi networks connecting devices can monitor and record data (including personal information). Such networks can only be secured by using other means of protection, such as a VPN or secure Hypertext Transfer Protocol over Transport Layer Security (HTTPS).

Wi-Fi Protected Access encryption (WPA2) is considered secure, provided a strong passphrase is used. A proposed modification to WPA2 is WPA-OTP or WPA3, which stores an on-chip optically generated onetime pad on all connected devices which is periodically updated via strong encryption then hashed with the data to be sent or received. This would be unbreakable using any (even quantum) computer system as the hashed data is essentially random and no pattern can be detected if it is implemented properly. Main disadvantage is that it would need multi-GB storage chips so would be expensive for the consumers.

Piggybacking

Piggybacking refers to access to a wireless Internet connection by bringing one's own computer within the range of another's wireless connection, and using that service without the subscriber's explicit permission or knowledge.

During the early popular adoption of 802.11, providing open access points for anyone within range to use was encouraged to cultivate wireless community networks, particularly since people on average use only a fraction of their downstream bandwidth at any given time.

Recreational logging and mapping of other people's access points has become known as wardriving. Indeed, many access points are intentionally installed without security turned on so that they can be used as a free service. Providing access to one's Internet connection in this fashion may breach the Terms of Service or contract with the ISP. These activities do not result in sanctions in most jurisdictions; however, legislation and case law differ considerably across the world. A proposal to leave graffiti describing available services was called warchalking. A Florida court case determined that owner laziness was not to be a valid excuse.

Piggybacking often occurs unintentionally – a technically unfamiliar user might not change the default "unsecured" settings to their access point and operating systems can be configured to connect automatically to any available wireless network. A user who happens to start up a laptop in the vicinity of an access point may find the computer has joined the network without any visible indication. Moreover, a user intending to join one network may instead end up on another one if the latter has a stronger signal. In combination with automatic discovery of other network resources this could possibly lead wireless users to send sensitive data to the wrong middle-man when seeking a destination. For example, a user could inadvertently use an unsecure network to log into a website, thereby making the login credentials available to anyone listening, if the website uses an unsecure protocol such as plain HTTP without TLS (HTTPS).

An unauthorized user can obtain security information (factory preset passphrase and/ or Wi-Fi Protected Setup PIN) from a label on a wireless access point can use this information (or connect by the Wi-Fi Protected Setup pushbutton method) to commit unauthorized and/or unlawful activities.

Health Concerns

The World Health Organization (WHO) says "no health effects are expected from exposure to RF fields from base stations and wireless networks", but notes that they promote research into effects from other RF sources. Although the WHO's International Agency for Research on Cancer (IARC) later classified radiofrequency electromagnetic fields as "possibly carcinogenic to humans (Group 2B)" (a category used when "a causal association is considered credible, but when chance, bias or confounding cannot be ruled out with reasonable confidence"), this was based on risks associated with wireless phone use rather than Wi-Fi networks.

The United Kingdom's Health Protection Agency reported in 2007 that exposure to Wi-Fi for a year results in the "same amount of radiation from a 20-minute mobile phone call".

A review of studies involving 725 people who claimed electromagnetic hypersensitivity, "...suggests that 'electromagnetic hypersensitivity' is unrelated to the presence of EMF, although more research into this phenomenon is required."

Super Wi-Fi

Super Wi-Fi is a term coined by the United States Federal Communications Commission (FCC) to describe a wireless networking proposal which the FCC plans to use for the creation of longer-distance wireless Internet access. The use of the trademark "Wi-Fi" in the name has been criticized because it is not based on Wi-Fi technology or endorsed by the Wi-Fi Alliance. A trade show has also been called the "Super WiFi Summit" (without hyphen). Various standards such as IEEE 802.22 and IEEE 802.11af have been proposed for this concept.

Instead of using the 2.4 GHz radio frequency of Wi-Fi, the "Super Wi-Fi" proposal uses the lower-frequency white spaces between television channel frequencies. These lower frequencies allow the signal to travel further and penetrate walls better than the higher frequencies previously used. The FCC's plan was to allow those white space frequencies to be used for free, as happens with shorter-range Wi-Fi and Bluetooth. However, due to concerns of interference to broadcast, Super Wi-Fi devices cannot access the TV spectrum at will. The FCC has made mandatory the utilization of a TV white space database (also referred to as geolocation database), which must be accessed by the Super Wi-Fi devices before the latter gain access to the VHF-UHF spectrum. The white space database evaluates the potential for interference to broadcast and either grant or deny access of Super Wi-Fi devices to the VHF-UHF spectrum.

On April 19, 2011, Rice University, in partnership with the nonprofit organization Technology For All, installed the first residential deployment of Super Wi-Fi in east Houston. The network uses white spaces for backhaul and provides access to clients using 2.4 GHz Wi-Fi.

On May 8, 2011, a public Super Wi-Fi network was developed in Calgary, Alberta. Calgary based company WestNet Wireless. launched the network for free and paid subscribers.

On January 26, 2012, the United States first public Super Wi-Fi network was developed in Wilmington, North Carolina. Florida based company Spectrum Bridge, Inc. launched the network for public use with access at Hugh MacRae park.

On July 9, 2013, West Virginia University launched the first campus Super WiFi network.

Wi-Fi Direct

Wi-Fi Direct, initially called Wi-Fi P2P, is a Wi-Fi standard enabling devices to easily connect with each other without requiring a wireless access point. Wi-Fi Direct allows two devices to establish a direct Wi-Fi connection without requiring a wireless router. Hence, Wi-Fi Direct is single radio hop communication, not multihop wireless communications, unlike wireless ad hoc networks and mobile ad hoc networks. Wi-Fi ad hoc mode, however, supports multi-hop radio communications, with intermediate Wi-Fi nodes as packet relays.

Wi-Fi becomes a way of communicating wirelessly, much like Bluetooth. It is useful for everything from internet browsing to file transfer, and to communicate with one or more devices simultaneously at typical Wi-Fi speeds. One advantage of Wi-Fi Direct is the ability to connect devices even if they are from different manufacturers. Only one of the Wi-Fi devices needs to be compliant with Wi-Fi Direct to establish a peer-to-peer connection that transfers data directly between them with greatly reduced setup.

Wi-Fi Direct negotiates the link with a Wi-Fi Protected Setup system that assigns each device a limited wireless access point. The "pairing" of Wi-Fi Direct devices can be set up to require the proximity of a near field communication, a Bluetooth signal, or a button press on one or all the devices.

Background

Basic Wi-Fi

Conventional Wi-Fi networks are typically based on the presence of controller devices known as wireless access points. These devices normally combine three primary functions:

- Physical support for wireless and wired networking

- Bridging and routing between devices on the network

- Service provisioning to add and remove devices from the network.

A typical Wi-Fi home network includes laptops, tablets and phones, devices like modern printers, music devices and televisions. The majority of Wi-Fi networks are set up in "infrastructure mode", where the access point acts as a central hub to which Wi-Fi capable devices are connected. The devices do not communicate directly with each other (that is, in "ad-hoc mode"), but they go through the access point. Wi-Fi Direct devices are able to communicate with each other without requiring a dedicated wireless access point. The Wi-Fi Direct devices negotiate when they first connect to determine which device shall act as an access point.

Automated Setup

With the increase in the number and type of devices attaching to Wi-Fi systems, the basic model of a simple router with smart computers became increasingly strained. At the same time, the increasing sophistication of the hot spots presented setup problems for the users. To address these problems, there have been numerous attempts to simplify certain aspects of the setup task.

A common example is the Wi-Fi Protected Setup system included in most access points built since 2007 when the standard was introduced. Wi-Fi Protected Setup allows access points to be set up simply by entering a PIN or other identification into a connection screen, or in some cases, simply by pressing a button. The Protected Setup system uses this information to send data to a computer, handing it the information needed to complete the network setup and connect to the Internet. From the user's point of view, a single click replaces the multi-step, jargon-filled setup experience formerly required.

While the Protected Setup model works as intended, it was intended only to simplify the connection between the access point and the devices that would make use of its services, primarily accessing the Internet. It provides little help *within* a network - finding and setting up printer access from a computer for instance. To address those roles, a number of different protocols have developed, including Universal Plug and Play (UPnP), Devices Profile for Web Services (DPWS), and Zero Configuration Networking (ZeroConf). These protocols allow devices to seek out other devices within the network, query their capabilities, and provide some level of automatic setup.

New Uses

Wi-Fi has become a standard feature in smart phones and portable media players, and in feature phones as well. The process of adding Wi-Fi to smaller devices has accelerated, and it is now possible to find printers, cameras, scanners and many other common devices with Wi-Fi in addition to other connections, like USB.

The widespread adoption of Wi-Fi in new classes of smaller devices made the need for ad hoc networking much more important. Even without a central Wi-Fi hub or router, it would be useful for a laptop computer to be able to wirelessly connect to a local printer. Although the ad hoc mode was created to address this sort of need, the lack of additional information for discovery makes it difficult to use in practice.

Although systems like UPnP and Bonjour provide many of the needed capabilities and are included in some devices, a single widely supported standard was lacking, and support within existing devices was far from universal. A guest using their smart phone would likely be able to find a hot spot and connect to the Internet with ease, perhaps using Protected Setup to do so. But the same device would find streaming music to a computer or printing a file might be difficult, or simply not supported between differing brands of hardware.

Wi-Fi Direct can provide a wireless connection to peripherals. Wireless mice, keyboards, remote controls, headsets, speakers, displays and many other functions can be implemented with Wi-Fi Direct. This has begun with Wi-Fi mouse products, and Wi-Fi Direct remote controls that were shipping circa November 2012.

File sharing applications on Android and BlackBerry 10 devices could use Wi-Fi Direct, with most Android Version 4.1 (Jellybean), introduced in July 2012, and BlackBerry

10.2 supported. Android version 4.2 (Jellybean) included further refinements to Wi-Fi Direct including persistent permissions enabling two-way transfer of data between multiple devices.

The Miracast standard for the wireless connection of devices to displays is based on Wi-Fi direct.

Technical Description

Wi-Fi Direct essentially embeds a software access point ("Soft AP"), into any device that must support Direct. The soft AP provides a version of Wi-Fi Protected Setup with its push-button or PIN-based setup.

When a device enters the range of the Wi-Fi Direct host, it can connect to it, and then gather setup information using a Protected Setup-style transfer. Connection and setup is so simplified that it may replace Bluetooth in some situations.

Soft APs can be as simple or as complex as the role requires. A digital picture frame might provide only the most basic services needed to allow digital cameras to connect and upload images. A smart phone that allows data tethering might run a more complex soft AP that adds the ability to bridge to the Internet. The standard also includes WPA2 security and features to control access within corporate networks. Wi-Fi Direct-certified devices can connect one-to-one or one-to-many and not all connected products need to be Wi-Fi Direct-certified. One Wi-Fi Direct enabled device can connect to legacy Wi-Fi certified devices.

The Wi-Fi Direct certification program is developed and administered by the Wi-Fi Alliance, the industry group that owns the "Wi-Fi" trademark. The specification is available for purchase from the Wi-Fi Alliance.

Commercialization

Laptops

Intel included Wi-Fi Direct on the Centrino 2 platform, in its My WiFi technology by 2008. Wi-Fi Direct devices can connect to a notebook computer that plays the role of a software Access Point (AP). The notebook computer can then provide Internet access to the Wi-Fi Direct-enabled devices without a Wi-Fi AP. Marvell Technology Group, Atheros, Broadcom, Intel, Ralink and Realtek announced their first products in October 2010. Redpine Signals's chipset was Wi-Fi Direct certified in November of the same year.

Mobile Devices

Google announced Wi-Fi Direct support in Android 4.0 in October 2011. While some Android 2.3 devices like Samsung Galaxy S II have had this feature through proprietary operating system extensions developed by OEMs, the Galaxy Nexus (released November 2011) was the first Android device to ship with Google's implementation of this feature and an application programming interface for developers. Ozmo Devices, which developed integrated circuits (chips) designed for Wi-Fi Direct, was acquired by Atmel in 2012.

Wi-Fi Direct became available with the Blackberry 10.2 upgrade.

By March 2016, no iPhone device implements Wi-Fi Direct; instead, iOS has its own proprietary feature.

An icon to indicate Wi-Fi Direct operation under Android.

Game Consoles

The Xbox One, released in 2013, supports Wi-Fi Direct.

NVIDIA's SHIELD controller uses Wi-Fi Direct to connect to compatible devices. NVIDIA claims a reduction in latency and increase in throughput over competing Bluetooth controllers.

Televisions

In March 2016 Sony, LG, Philips and X.VISION have implemented Wi-Fi Direct on some of their televisions.

Long-range Wi-Fi

Long-range Wi-Fi is used for low-cost, unregulated point-to-point computer network connections, as an alternative to other fixed wireless, cellular networks or satellite Internet access.

Wi-Fi networks have a range that's limited by the transmission power, antenna type, the location they're used in, and the environment. A typical wireless router in an indoor point-to-multipoint arrangement using 802.11n and a stock antenna might have a range of 32 metres (105 ft). Outdoor point-to-point arrangements, through use of directional antennas, can be extended with many kilometers between stations.

Since the development of the IEEE 802.11 radio standard (marketed under the Wi-Fi brand name), the technology has become markedly less expensive and achieved higher bit rates. Long-range Wi-Fi especially in the 2.4 GHz band (as the shorter-range higher-bit-rate 5.8 GHz bands become popular alternatives to wired LAN connections) have proliferated with specialist devices. While Wi-Fi hotspots are ubiquitous in urban areas, some rural areas use more powerful longer-range transceivers as alternatives to cell (GSM, CDMA) or fixed wireless (Motorola Canopy and other 900 MHz) applications. The main drawbacks of 2.4 GHz vs. these lower-frequency options are:

- poor signal penetration – 2.4 GHz connections are effectively limited to line of sight or soft obstacles;

- far less range – GSM or CDMA cell phones can connect reliably at > 16 km (9.9 mi) distances; the range of GSM, imposed by the parameters of time-division multiple access, is set at 35 km;

- few service providers commercially support long-distance Wi-Fi connections.

Despite a lack of commercial service providers, applications for long-range Wi-Fi have cropped up around the world. It has also been used in experimental trials in the developing world to link communities separated by difficult geography with few or no other connectivity options. Some benefits of using long-range Wi-Fi for these applications include:

- unlicensed spectrum – avoiding negotiations with incumbent telecom providers, governments or others;

- smaller, simpler, cheaper antennas – 2.4 GHz antennas have less than half the size of comparable-strength 900 MHz antennas and require less lightning protection;

- availability of proven free software like OpenWrt, DD-WRT, Tomato that works even on old routers (WRT54G, for instance) and makes modes like WDS, OLSR, etc., available to anyone, including revenue-sharing models for hotspots.

Nonprofit organizations operating widespread installations, such as forest services, also make extensive use of long-range Wi-Fi to augment or replace older communications technologies such as shortwave or microwave transceivers in licensed bands.

Applications

Business

- Provide coverage to a large office or business complex or campus.

- Establish point-to-point link between large skyscrapers or other office buildings or airports.

- Bring Internet to remote construction sites or research labs.

- Simplify networking technologies by coalescing around a small number of Internet related widely used technologies, limiting or eliminating legacy technologies such as shortwave radio so these can be dedicated to uses where they actually are needed.

- Bring Internet to a home if regular cable/DSL cannot be hooked up at the location.

- Bring Internet to a vacation home or cottage on a remote mountain or on a lake.

- Bring Internet to a yacht or large seafaring vessel.

- Share a neighborhood Wi-Fi network.

Nonprofit and Government

- Connect widespread physical guard posts, e.g. for foresters, that guard a physical area, without any new wiring

- In tourist regions, fill in cell dead zones with Wi-Fi coverage, and ensure connectivity for local tourist trade operators

- Reduce costs of dedicated network infrastructure and improve security by applying modern encryption and authentication.

Military

- Connect critical opinion leaders, infrastructure such as schools and police stations, in a network local authorities can maintain

- Build resilient infrastructure with cheaper equipment that an impoverished war-torn region can afford, i.e. using commercial grade, rather than military-class network technology, which may then be left with the developed-world military

- Reduce costs and simplify/protect supply chains by using cheaper simpler equipment that draws less fuel and battery power; *In general these are high priorities for commercial technologies like Wi-Fi especially as they are used in mobile devices.*

Scientific Research

- A long-range seismic sensor network was used during the Andean Seismic Project in Peru. A multi-hop span with a total length of 320 kilometres was crossed with some segments around 30 to 50 kilometers. The goal was to connect to outlying stations to UCLA in order to receive seismic data in real time.

Large-scale Deployments

The Technology and Infrastructure for Emerging Regions (TIER) project at University of California at Berkeley in collaboration with Intel, uses a modified Wi-Fi setup to create long-distance point-to-point links for several of its projects in the developing world. This technique, dubbed Wi-Fi over Long Distance (WiLD), is used to connect the Aravind Eye Hospital with several outlying clinics in Tamil Nadu state, India. Distances range from five to over fifteen kilometres (3–10 miles) with stations placed in line of sight of each other. These links allow specialists at the hospital to communicate with nurses and patients at the clinics through video conferencing. If the patient needs further examination or care, a hospital appointment can then be scheduled. Another network in Ghana links the University of Ghana, Legon campus to its remote campuses at the Korle bu Medical School and the City campus; a further extension will feature links up to 80 km (50 mi) apart.

The Tegola project of the University of Edinburgh is developing new technologies to bring high-speed, affordable broadband to rural areas beyond the reach of fibre. A 5-link ring connects Knoydart, the N. shore of Loch Hourne, and a remote community at Kilbeg to backhaul from the Gaelic College on Skye. All links pass over tidal waters; they range in length from 2.5 km to 19 km.

Increasing Range in Other Ways

Specialized Wi-Fi Channels

In most standard Wi-Fi routers, the three standards, a, b and g, are enough. But in long-range Wi-Fi, special technologies are used to get the most out of a Wi-Fi connection. The 802.11-2007 standard adds 10 MHz and 5 MHz OFDM modes to the 802.11a standard, and extend the time of cyclic prefix protection from 0.8 μs to 3.2 μs, quadrupling the multipath distortion protection. Some commonly available 802.11a/g chipsets support the OFDM 'half-clocking' and 'quarter-clocking' that is in the 2007 standard, and 4.9 GHz and 5.0 GHz products are available with 10 MHz and 5 MHz channel bandwidths. It is likely that some 802.11n D.20 chipsets will also support 'half-clocking' for use in 10 MHz channel bandwidths, and at double the range of the 802.11n standard.

802.11n and MIMO

Preliminary 802.11n working became available in many routers in 2008. This technology can use multiple antennas to target one or more sources to increase speed. This is known as MIMO, Multiple Input Multiple Output. In tests, the speed increase was said to only occur over short distances rather than the long range needed for most point-to-point setups. On the other hand, using dual antennas with orthogonal polarities along with a 2x2 MIMO chipset effectively enable two independent carrier signals to be sent and received along the same long distance path.

Power Increase or Receiver Sensitivity Boosting

A rooftop 1 watt Wi-Fi amp, feeding a simple vertical antenna on the left.

Another way of adding range uses a power amplifier. Commonly known as "range extender amplifiers" these small devices usually supply around ½ watt of power to the antenna. Such amplifiers may give more than five times the range to an existing network. Every 3 dB gain doubles the effective output power. An antenna receiving 1 watt of power, and 6 dB gain would have an effective power of 4 watts. The alternative techniques of selecting a more sensitive WLAN adapter and more directive antenna should also be considered.

Higher Gain Antennas and Adapter Placement

Specially shaped directional antennas can increase the range of a Wi-Fi transmission without a drastic increase in transmission power. High gain antenna may be of many designs, but all allow transmitting a narrow signal beam over greater distance than a non-directional antenna, often nulling out nearby interference sources. A popular low-cost home made approach increases WiFi ranges by just placing standard USB WLAN hardware at the focal point of modified parabolic cookware. Such "WokFi" techniques typically yield gains more than 10 dB over the bare system; enough for line of sight (LOS) ranges of several kilometers and improvements in marginal locations. Although often low power, cheap USB WLAN adapters suit site auditing and location of local signal "sweet spots". As USB leads incur none of the losses normally associated with costly microwave coax and SMA fittings, just extending a USB adapter (or AP, etc.) up to a window, or away from shielding metal work and vegetation, may dramatically improve the link.

Protocol Hacking

The standard IEEE 802.11 protocol implementations can be modified to make them more suitable for long distance, point-to-point usage, at the risk of breaking interoperability with other Wi-Fi devices and suffering interference from transmitters located near the antenna. These approaches are used by the TIER project.

In addition to power levels, it is also important to know how the 802.11 protocol acknowledges each received frame. If the acknowledgement is not received, the frame is re-transmitted. By default, the maximum distance between transmitter and receiver is 1.6 km (1 mi). On longer distances the delay will force retransmissions. On standard firmware for some professional equipment such as the Cisco Aironet 1200, this parameter can be tuned for optimal throughput. OpenWrt, DD-WRT and all derivatives of it also enable such tweaking. In general, open source software is vastly superior to commercial firmware for all purposes involving protocol hacking, as the philosophy is to expose all radio chipset capabilities and let the user modify them. This strategy has been especially effective with low end routers such as the WRT54G which had excellent hardware features the commercial firmware did not support. As of 2011, many vendors still supported only a subset of chipset features that open source firmware unlocked, and most vendors actively encourage the use of open source firmware for protocol hacking, in part to avoid the difficulty of trying to support commercial firmware users attempting this.

Packet fragmentation can also be used to improve throughput in noisy/congested conditions. Although packet fragmentation is often thought of as something bad, and does indeed add a large overhead, reducing throughput, it is sometimes necessary. For example, in a congested situation, ping times of 30 byte packets can be excellent, while ping times of 1450 byte packets can be very poor with high packet loss. Dividing the packet in half, by setting the fragmentation threshold to 750, can vastly improve the throughput. The fragmentation threshold should be a division of the MTU, typically 1500, so should be 750, 500, 375, etc. However, excessive fragmentation can make the problem worse, since the increased overhead will increase congestion.

Obstacles to Long-range Wi-Fi

Methods that increase the range of a Wi-Fi connection may also make it fragile and volatile, due to various factors including:

Landscape Interference

Obstacles are among the biggest problems when setting up a long-range Wi-Fi. Trees and forests attenuate the microwave signal, and hills make it difficult to establish line-of-sight propagation.

In a city, buildings will impact integrity, speed and connectivity. Steel frames and Sheet metal in walls or roofs may partially or fully reflect radio signals, causing signal loss or multipath problems. Concrete or plaster walls absorb microwave signals significantly, reducing the total signal.

Tidal Fading

When point-to-point wireless connections cross tidal estuaries or archipelagos, multipath interference from reflections over tidal water can be considerably destructive. The Tegola project uses a slow frequency-hopping technique to mitigate tidal fading.

2.4 GHz Interference

Microwave ovens in residences dominate the 2.4 GHz band and will cause "meal time perturbations" of the noise floor. There are many other sources of interference that aggregate into a formidable obstacle to enabling long-range use in occupied areas. Residential wireless phones, baby monitors, wireless cameras, remote car starters, and Bluetooth products are all capable of transmitting in the 2.4 GHz band.

Due to the intended nature of the 2.4 GHz band, there are many users of this band, with potentially dozens of devices per household. By its very nature, "long range" connotes an antenna system which can see many of these devices, which when added together produce a very high noise floor, whereby no single signal is usable, but nonetheless are still received. The aim of a long-range system is to produce a system which over-powers these signals and/or uses directional antennas to prevent the receiver "seeing" these devices, thereby reducing the noise floor.

Notable Links

Italy

The longest unamplified Wi-Fi link is a 304 km link achieved by CISAR (Italian Center for Radio Activities).

- link first established on 2007-06-16
- it appears to be permanent from Monte Amiata (Tuscany) to Monte Limbara (Sardinia)
- frequency: 5765 MHz
- IEEE 802.11ac (Wi-Fi), bandwidth 5 MHz
- Radio: Ubiquiti Networks XR5
- Wireless routers: MikroTik RouterBOARD with RouterOS, NStreme optimization enabled
- Length: 304 km (189 mi).
- Antenna is 120 cm with handmade waveguide. 35 dBi estimated

New world record for long-range wireless broadband

- link first established on 2016-05-7 e 2016-05-8

- it appears to be permanent from Monte Amiata (Tuscany) to Monte Limbara (Sardinia)

- frequency: 5765 MHz

- IEEE 802.11a (Wi-Fi), bandwidth 50 MHz

- data rates: of up to 356.33 Mbit/s

- Radio: Ubiquiti Networks AF-5X radios

- Wireless routers: Ubiquiti airFiber®

- Length: 304 km (189 mi).

- Antenna is 120 cm with handmade waveguide. 35 dBi estimated

Venezuela

Another notable unamplified Wi-Fi link is a 279 km link achieved by the Latin American Networking School Foundation.

- Pico del Águila - El Baúl Link.

- frequency: 2412 MHz

- link established in 2006

- IEEE 802.11 (Wi-Fi), channel 1, bandwidth 22 MHz

- Wireless routers: Linksys WRT54G, OpenWrt firmware at el Águila and DD-WRT firmware at El Baúl.

- Length: 279 km (173 mi).

- Parabolic dish antennas were used at both ends, recycled from satellite service.

- At El Aguila site an aluminum mesh reflector 2.74 m (9 ft) diameter, center-fed, at El Baúl a fiberglass solid reflector, offset-fed, 2.44 by 2.74 m (8 by 9 ft). On both ends the feeds were 12 dBi Yagis.

- Linksys WRT54G series routers fed the antennas with short LMR400 cables, so the effective gain of the complete antenna is estimated at about 30 dBi.

- This is the largest known range attained with this technology, improving on a previous US record of 201 km (125 mi) achieved last year in U.S. The Swedish space agency attained 315 km (196 mi), but using 6 watt amplifiers to reach an overhead stratospheric balloon.

Peru

Antenna's installation at Napo, Loreto (March 2007)

Loreto, in the jungle region of Peru, is the location of the longest Wi-Fi-based multihop network in the world. This network has been implemented by the Rural Telecommunications Research Group of the Pontificia Universidad Católica del Perú (GTR PUCP). The Wi-Fi chain goes through many small villages and takes seventeen hops to cover the whole distance. It begins in Cabo Pantoja's Health Post and finishes at downtown Iquitos. Its length is about 445 km. The intervention zone was established in the lowland jungle with elevations under 500 meters above sea level. It is a flat zone and for this reason GTR PUCP installed towers with an average height of 80 meters.

- The link was established in 2007. GTR PUCP, the regional government of Loreto, and Vicariate San José de Amazonas are working together on maintenance of the network.

- Frequency channels used: 1, 6 and 11, 802.11g non-interfered channels

- smartBridges Wireless Routers were used.

- L-com antennas were used.

Hotspot (Wi-Fi)

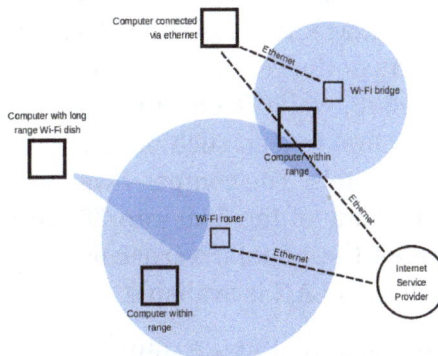

A diagram showing a Wi-Fi network

A hotspot is a physical location where people may obtain Internet access, typically using Wi-Fi technology, via a wireless local area network (WLAN) using a router connected to an internet service provider.

Public hotspots may be found in a number of businesses for use of customers in many developed urban areas throughout the world, such as coffee shops or hotels. Public hotspots are typically built from wireless access points configured to provide Internet access, controlled to some degree by the venue. Private hotspots may be configured on a smart phone or tablet with a mobile network data plans to allow Internet access to a few other devices via WiFi.

History

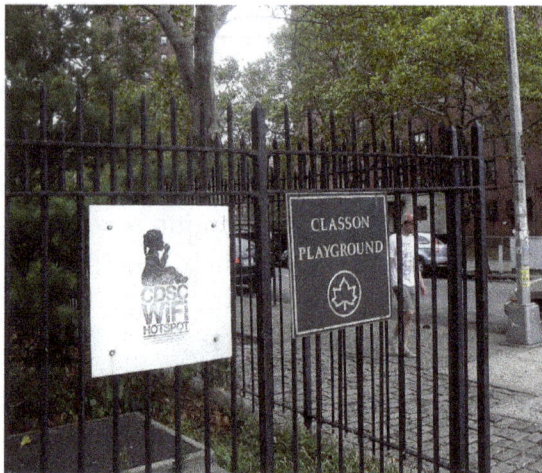

Public park in Brooklyn, New York, has free Wi-Fi from a local corporation

Public access wireless local area networks (LANs) were first proposed by Henrik Sjödin at the NetWorld+Interop conference in The Moscone Center in San Francisco in August 1993. Sjödin did not use the term hotspot but referred to publicly accessible wireless LANs.

The first commercial venture to attempt to create a public local area access network was a firm founded in Richardson, Texas known as PLANCOM (Public Local Area Network Communications). The founders of the venture, Mark Goode, Greg Jackson, and Brett Stewart dissolved the firm in 1998, while Goode and Jackson created MobileStar Networks. The firm was one of the first to sign such public access locations as Starbucks, American Airlines, and Hilton Hotels. The company was sold to Deutsche Telecom in 2001, who then converted the name of the firm into "T-Mobile Hotspot." It was then that the term "hotspot" entered the popular vernacular as a reference to a location where a publicly accessible wireless LAN is available.

ABI Research reported there was a total of 4.9 million global Wi-Fi hotspots in 2012 and projected that number would surpass 6.3 million by the end of 2013. The latest

Wireless Broadband Alliance (WBA) Industry Report outlines a positive scenario for the Wi-Fi market: a steady annual increase from 5.2m public hotspots in 2012 to 10.5m public hotspots in 2018. Collectively, WBA operator members serve more than 1 billion subscribers and operate more than 15 million hotspots globally.

Uses

The public can use a laptop or other suitable portable device to access the wireless connection (usually Wi-Fi) provided. Of the estimated 150 million laptops, 14 million PDAs, and other emerging Wi-Fi devices sold per year for the last few years, most include the Wi-Fi feature.

For venues that have broadband Internet access, offering wireless access is as simple as configuring one access point (AP), in conjunction with a router and connecting the AP to the Internet connection. A single wireless router combining these functions may suffice.

The iPass 2014 interactive map, that shows data provided by the analysts Maravedis Rethink, shows that in December 2014 there are 46,000,000 hotspots worldwide and more than 22,000,000 roamable hotspots. More than 10,900 hotspots are on trains, planes and airports (Wi-Fi in motion) and more than 8,500,000 are "branded" hotspots (retail, cafés, hotels). The region with the largest number of public hotspots is Europe, followed by North America and Asia.

Security

Security is a serious concern in connection with Hotspots. There are three possible attack vectors. First, there is the wireless connection between the client and the access point. This needs to be encrypted, so that the connection cannot be eavesdropped or attacked by a man-in-the-middle-attack. Second, there is the Hotspot itself. The WLAN encryption ends at the interface, then travels its network stack unencrypted and then, third, travels over the wired connection up to the BRAS of the ISP.

The safest method when accessing the Internet over a Hotspot, with unknown security measures, is end-to-end encryption. Examples of strong end-to-end encryption are HTTPS and SSH.

Locations

Hotspots are often found at airports, bookstores, coffee shops, department stores, fuel stations, hotels, hospitals, libraries, public pay phones, restaurants, RV parks and campgrounds, supermarkets, train stations, and other public places. Additionally, many schools and universities have wireless networks in their campuses.

Types

Free hotspots operate in two ways:

- Using an open public network is the easiest way to create a free hotspot. All that is needed is a Wi-Fi router. Similarly, when users of private wireless routers turn off their authentication requirements, opening their connection, intentionally or not, they permit piggybacking (sharing) by anyone in range.

- Closed public networks use a HotSpot Management System to control access to hotspots. This software runs on the router itself or an external computer allowing operators to authorize only specific users to access the Internet. Providers of such hotspots often associate the free access with a menu, membership, or purchase limit. Operators may also limit each user's available bandwidth (upload and download speed) to ensure that everyone gets a good quality service. Often this is done through service-level agreements.

Commercial Hotspots

A commercial hotspot may feature:

- A captive portal / login screen / splash page that users are redirected to for authentication and/or payment. The captive portal / splash page sometimes includes the social login buttons.

- A payment option using a credit card, iPass, PayPal, or another payment service (voucher-based Wi-Fi)

- A walled garden feature that allows free access to certain sites

- Service-oriented provisioning to allow for improved revenue

- Data analytics and data capture tools, to analyze and export data from Wi-Fi clients

Many services provide payment services to hotspot providers, for a monthly fee or commission from the end-user income. For example, Amazingports can be used to set up hotspots that intend to offer both fee-based and free internet access, and ZoneCD is a Linux distribution that provides payment services for hotspot providers who wish to deploy their own service.

Major airports and business hotels are more likely to charge for service, though most hotels provide free service to guests; and increasingly, small airports and airline lounges offer free service.. Retail shops, public venues and offices usually provide a free Wi-Fi SSID for their guests and visitors.

Roaming services are expanding among major hotspot service providers. With roaming service the users of a commercial provider can have access to other providers' hotspots, either free of charge or for extra fees, which users will usually be charged on an access-per-minute basis.

Software Hotspots

Many Wi-Fi adapters built into or easily added to consumer computers and mobile devices include the functionality to operate as private or mobile hotspots, sometimes referred to as "mi-fi". The use of a private hotspot to enable other personal devices to access the WAN (usually but not always the Internet) is a form of bridging, and known as tethering. Manufacturers and firmware creators can enable this functionality in Wi-Fi devices on many Wi-Fi devices, depending upon the capabilities of the hardware, and most modern consumer operating systems, including Android, Apple OS X 10.6 and later, Windows mobile, and Linux include features to support this. Additionally wireless chipset manufacturers such as Atheros, Broadcom, Intel and others, may add the capability for certain Wi-Fi NICs, usually used in a client role, to also be used for hotspot purposes. However, some service providers, such as AT&T, Sprint, and T-Mobile charge users for this service or prohibit and disconnect user connections if tethering is detected.

Third-party software vendors offer applications to allow users to operate their own hotspot, whether to access the Internet when on the go, share an existing connection, or extend the range of another hotspot.

Hotspot 2.0

Hotspot 2.0, also known as HS2 and Wi-Fi Certified Passpoint, is an approach to public access Wi-Fi by the Wi-Fi Alliance. The idea is for mobile devices to automatically join a Wi-Fi subscriber service whenever the user enters a Hotspot 2.0 area, in order to provide better bandwidth and services-on-demand to end-users and relieve carrier infrastructure of some traffic.

Hotspot 2.0 is based on the IEEE 802.11u standard, which is a set of protocols published in 2011 to enable cellular-like roaming. If the device supports 802.11u and is subscribed to a Hotspot 2.0 service it will automatically connect and roam.

Supported Devices

- Some Chinese tablet computers

- Some THL smartphones

- Apple mobile devices running iOS 7 and up

- Some Samsung Galaxy smartphones

- Windows 10 devices have full support for network discovery and connection

- Windows 8 and Windows 8.1 lack network discovery, but support connecting to a network when the credentials are known

Billing

EDCF User-Priority-List							
		Net traffic					
		low			high		
		Audio	Video	Data	Audio	Video	Data
User needs	time-critical	7	5	0	6	4	0
	not time-critical	-	-	2	-	-	2

The so-called "User-Fairness-Model" is a dynamic billing model, which allows volume-based billing, charged only by the amount of payload (data, video, audio). Moreover, the tariff is classified by net traffic and user needs.

If the net traffic increases, then the user has to pay the next higher tariff class. The user can be prompted to confirm that they want to continue the session in the higher traffic class. A higher class fare can also be charged for delay sensitive applications such as video and audio, versus non time-critical applications such as reading Web pages and sending e-mail.

Tariff Classes of the User-Fairness-Model			
		Net traffic	
		low	high
User needs	time-critical	standard	exclusive
	not time-critical	low priced	standard

The "User-fairness model" can be implemented with the help of EDCF (IEEE 802.11e). A EDCF user priority list shares the traffic in 3 access categories (data, video, audio) and user priorities (UP).

- Data [UP 0|2]

- Video [UP 5|4]

- Audio [UP 7|6]

Security Concerns

Some hotspots authenticate users; however, this does not prevent users from viewing network traffic using packet sniffers.

Some vendors provide a download option that deploys WPA support. This conflicts with enterprise configurations that have solutions specific to their internal WLAN.

In order to provide robust security to hotspot users, the Wi-Fi Alliance is developing a new hotspot program that aims to encrypt hotspot traffic with WPA2 security. The program was scheduled to launch in the first half of 2012.

Legal Concerns

Depending upon the location, providers of public hotspot access may have legal obligations, related to privacy requirements and liability for use for unlawful purposes. In countries where the internet is regulated or freedom of speech more restricted, there may be requirements such as licensing, logging, or recording of user information. Concerns may also relate to child safety, and social issues such as exposure to objectionable content, protection against cyberbullying and illegal behaviours, and prevention of perpetration of such behaviors by hotspot users themselves.

European Union

- Data Retention Directive Hotspot owners must retain key user statistics for 12 months.

- Directive on Privacy and Electronic Communications

United Kingdom

- Data Protection Act 1998 The hotspot owner must retain individual's information within the confines of the law.

- Digital Economy Act 2010 Deals with, among other things, copyright infringement, and imposes fines of up to £250,000 for contravention.

IEEE 802.11

IEEE 802.11 is a set of media access control (MAC) and physical layer (PHY) specifications for implementing wireless local area network (WLAN) computer communication in the 900 MHz and 2.4, 3.6, 5, and 60 GHz frequency bands. They are created and maintained by the Institute of Electrical and Electronics Engineers (IEEE) LAN/MAN Standards Committee (IEEE 802). The base version of the standard was released in 1997, and has had subsequent amendments. The standard and amendments provide the basis for wireless network products using the Wi-Fi brand. While each amendment is officially revoked when it is incorporated in the latest version of the standard, the corporate world tends to market to the revisions because they concisely denote capabilities of their products. As a result, in the marketplace, each revision tends to become its own standard.

This Linksys WRT54GS WiFi router from 2005 operates on the 2.4 GHz "G" standard, capable of transmitting 54 megabits per second.

General Description

The 802.11 family consists of a series of half-duplex over-the-air modulation techniques that use the same basic protocol. 802.11-1997 was the first wireless networking standard in the family, but 802.11b was the first widely accepted one, followed by 802.11a, 802.11g, 802.11n, and 802.11ac. Other standards in the family (c–f, h, j) are service amendments that are used to extend the current scope of the existing standard, which may also include corrections to a previous specification.

802.11b and 802.11g use the 2.4 GHz ISM band, operating in the United States under Part 15 of the U.S. Federal Communications Commission Rules and Regulations. Because of this choice of frequency band, 802.11b and g equipment may occasionally suffer interference from microwave ovens, cordless telephones, and Bluetooth devices. 802.11b and 802.11g control their interference and susceptibility to interference by using direct-sequence spread spectrum (DSSS) and orthogonal frequency-division multiplexing (OFDM) signaling methods, respectively. 802.11a uses the 5 GHz U-NII band, which, for much of the world, offers at least 23 non-overlapping channels rather than the 2.4 GHz ISM frequency band offering only three non-overlapping channels, where other adjacent channels overlap. Better or worse performance with higher or lower frequencies (channels) may be realized, depending on the environment. 802.11n can use either the 2.4 GHz or the 5 GHz band; 802.11ac uses only the 5 GHz band.

The segment of the radio frequency spectrum used by 802.11 varies between countries. In the US, 802.11a and 802.11g devices may be operated without a license, as allowed in Part 15 of the FCC Rules and Regulations. Frequencies used by channels one through six of 802.11b and 802.11g fall within the 2.4 GHz amateur radio band. Licensed amateur radio operators may operate 802.11b/g devices under Part 97 of the FCC Rules and Regulations, allowing increased power output but not commercial content or encryption.

History

802.11 technology has its origins in a 1985 ruling by the U.S. Federal Communications Commission that released the ISM band for unlicensed use.

In 1991 NCR Corporation/AT&T (now Nokia Labs and LSI Corporation) invented a precursor to 802.11 in Nieuwegein, the Netherlands. The inventors initially intended to use the technology for cashier systems. The first wireless products were brought to the market under the name WaveLAN with raw data rates of 1 Mbit/s and 2 Mbit/s.

Vic Hayes, who held the chair of IEEE 802.11 for 10 years, and has been called the "father of Wi-Fi", was involved in designing the initial 802.11b and 802.11a standards within the IEEE.

In 1999, the Wi-Fi Alliance was formed as a trade association to hold the Wi-Fi trademark under which most products are sold.

Protocol

802.11-1997 (802.11 legacy)

The original version of the standard IEEE 802.11 was released in 1997 and clarified in 1999, but is now obsolete. It specified two net bit rates of 1 or 2 megabits per second (Mbit/s), plus forward error correction code. It specified three alternative physical layer technologies: diffuse infrared operating at 1 Mbit/s; frequency-hopping spread spectrum operating at 1 Mbit/s or 2 Mbit/s; and direct-sequence spread spectrum operating at 1 Mbit/s or 2 Mbit/s. The latter two radio technologies used microwave transmission over the Industrial Scientific Medical frequency band at 2.4 GHz. Some earlier WLAN technologies used lower frequencies, such as the U.S. 900 MHz ISM band.

Legacy 802.11 with direct-sequence spread spectrum was rapidly supplanted and popularized by 802.11b.

802.11a (OFDM Waveform)

Originally described as clause 17 of the 1999 specification, the OFDM waveform at 5.8 GHz is now defined in clause 18 of the 2012 specification, and provides protocols that allow transmission and reception of data at rates of 1.5 to 54 Mbit/s. It has seen widespread worldwide implementation, particularly within the corporate workspace. While the original amendment is no longer valid, the term *802.11a* is still used by wireless access point (cards and routers) manufacturers to describe interoperability of their systems at 5 GHz, 54 Mbit/s.

The 802.11a standard uses the same data link layer protocol and frame format as the original standard, but an OFDM based air interface (physical layer). It operates in the

5 GHz band with a maximum net data rate of 54 Mbit/s, plus error correction code, which yields realistic net achievable throughput in the mid-20 Mbit/s.

Since the 2.4 GHz band is heavily used to the point of being crowded, using the relatively unused 5 GHz band gives 802.11a a significant advantage. However, this high carrier frequency also brings a disadvantage: the effective overall range of 802.11a is less than that of 802.11b/g. In theory, 802.11a signals are absorbed more readily by walls and other solid objects in their path due to their smaller wavelength, and, as a result, cannot penetrate as far as those of 802.11b. In practice, 802.11b typically has a higher range at low speeds (802.11b will reduce speed to 5.5 Mbit/s or even 1 Mbit/s at low signal strengths). 802.11a also suffers from interference, but locally there may be fewer signals to interfere with, resulting in less interference and better throughput.

802.11b

The 802.11b standard has a maximum raw data rate of 11 Mbit/s, and uses the same media access method defined in the original standard. 802.11b products appeared on the market in early 2000, since 802.11b is a direct extension of the modulation technique defined in the original standard. The dramatic increase in throughput of 802.11b (compared to the original standard) along with simultaneous substantial price reductions led to the rapid acceptance of 802.11b as the definitive wireless LAN technology.

Devices using 802.11b experience interference from other products operating in the 2.4 GHz band. Devices operating in the 2.4 GHz range include microwave ovens, Bluetooth devices, baby monitors, cordless telephones, and some amateur radio equipment.

802.11g

In June 2003, a third modulation standard was ratified: 802.11g. This works in the 2.4 GHz band (like 802.11b), but uses the same OFDM based transmission scheme as 802.11a. It operates at a maximum physical layer bit rate of 54 Mbit/s exclusive of forward error correction codes, or about 22 Mbit/s average throughput. 802.11g hardware is fully backward compatible with 802.11b hardware, and therefore is encumbered with legacy issues that reduce throughput by ~21% when compared to 802.11a.

The then-proposed 802.11g standard was rapidly adopted in the market starting in January 2003, well before ratification, due to the desire for higher data rates as well as to reductions in manufacturing costs. By summer 2003, most dual-band 802.11a/b products became dual-band/tri-mode, supporting a and b/g in a single mobile adapter card or access point. Details of making b and g work well together occupied much of the lingering technical process; in an 802.11g network, however, activity of an 802.11b participant will reduce the data rate of the overall 802.11g network.

Like 802.11b, 802.11g devices suffer interference from other products operating in the 2.4 GHz band, for example wireless keyboards.

802.11-2007

In 2003, task group TGma was authorized to "roll up" many of the amendments to the 1999 version of the 802.11 standard. REVma or 802.11ma, as it was called, created a single document that merged 8 amendments (802.11a, b, d, e, g, h, i, j) with the base standard. Upon approval on March 8, 2007, 802.11REVma was renamed to the then-current base standard IEEE 802.11-2007.

802.11n

802.11n is an amendment that improves upon the previous 802.11 standards by adding multiple-input multiple-output antennas (MIMO). 802.11n operates on both the 2.4 GHz and the 5 GHz bands. Support for 5 GHz bands is optional. It operates at a maximum net data rate from 54 Mbit/s to 600 Mbit/s. The IEEE has approved the amendment, and it was published in October 2009. Prior to the final ratification, enterprises were already migrating to 802.11n networks based on the Wi-Fi Alliance's certification of products conforming to a 2007 draft of the 802.11n proposal.

802.11-2012

In May 2007, task group TGmb was authorized to "roll up" many of the amendments to the 2007 version of the 802.11 standard. REVmb or 802.11mb, as it was called, created a single document that merged ten amendments (802.11k, r, y, n, w, p, z, v, u, s) with the 2007 base standard. In addition much cleanup was done, including a reordering of many of the clauses. Upon publication on March 29, 2012, the new standard was referred to as IEEE 802.11-2012.

802.11ac

IEEE 802.11ac-2013 is an amendment to IEEE 802.11, published in December 2013, that builds on 802.11n. Changes compared to 802.11n include wider channels (80 or 160 MHz versus 40 MHz) in the 5 GHz band, more spatial streams (up to eight versus four), higher-order modulation (up to 256-QAM vs. 64-QAM), and the addition of Multi-user MIMO (MU-MIMO). As of October 2013, high-end implementations support 80 MHz channels, three spatial streams, and 256-QAM, yielding a data rate of up to 433.3 Mbit/s per spatial stream, 1300 Mbit/s total, in 80 MHz channels in the 5 GHz band. Vendors have announced plans to release so-called "Wave 2" devices with support for 160 MHz channels, four spatial streams, and MU-MIMO in 2014 and 2015.

802.11ad

IEEE 802.11ad is an amendment that defines a new physical layer for 802.11 networks to operate in the 60 GHz millimeter wave spectrum. This frequency band has significantly different propagation characteristics than the 2.4 GHz and 5 GHz bands where Wi-Fi networks operate. Products implementing the 802.11ad standard are being brought to market under the WiGig brand name. The certification program is now being developed by the Wi-Fi Alliance instead of the now defunct WiGig Alliance. The peak transmission rate of 802.11ad is 7 Gbit/s.

TP-Link announced the world's first 802.11ad router in January 2016.

802.11af

IEEE 802.11af, also referred to as "White-Fi" and "Super Wi-Fi", is an amendment, approved in February 2014, that allows WLAN operation in TV white space spectrum in the VHF and UHF bands between 54 and 790 MHz. It uses cognitive radio technology to transmit on unused TV channels, with the standard taking measures to limit interference for primary users, such as analog TV, digital TV, and wireless microphones. Access points and stations determine their position using a satellite positioning system such as GPS, and use the Internet to query a geolocation database (GDB) provided by a regional regulatory agency to discover what frequency channels are available for use at a given time and position. The physical layer uses OFDM and is based on 802.11ac. The propagation path loss as well as the attenuation by materials such as brick and concrete is lower in the UHF and VHF bands than in the 2.4 and 5 GHz bands, which increases the possible range. The frequency channels are 6 to 8 MHz wide, depending on the regulatory domain. Up to four channels may be bonded in either one or two contiguous blocks. MIMO operation is possible with up to four streams used for either space–time block code (STBC) or multi-user (MU) operation. The achievable data rate per spatial stream is 26.7 Mbit/s for 6 and 7 MHz channels, and 35.6 Mbit/s for 8 MHz channels. With four spatial streams and four bonded channels, the maximum data rate is 426.7 Mbit/s for 6 and 7 MHz channels and 568.9 Mbit/s for 8 MHz channels.

802.11ah

IEEE 802.11ah defines a WLAN system operating at sub-1 GHz license-exempt bands, with final approval slated for September 2016. Due to the favorable propagation characteristics of the low frequency spectra, 802.11ah can provide improved transmission range compared with the conventional 802.11 WLANs operating in the 2.4 GHz and 5 GHz bands. 802.11ah can be used for various purposes including large scale sensor networks, extended range hotspot, and outdoor Wi-Fi for cellular traffic offloading, whereas the available bandwidth is relatively narrow. The protocol intends consumption to be competitive with low power Bluetooth, at a much wider range.

802.11ai

IEEE 802.11ai is an amendment to the 802.11 standard that added new mechanisms for a faster initial link setup time.

802.11aj

IEEE 802.11aj is a rebanding of 802.11ad for use in the 45 GHz unlicensed spectrum available in some regions of the world (specifically China).

802.11aq

IEEE 802.11aq is an amendment to the 802.11 standard that will enable pre-association discovery of services. This extends some of the mechanisms in 802.11u that enabled device discovery to further discover the services running on a device, or provided by a network.

802.11ax

IEEE 802.11ax is the successor to 802.11ac, and will increase the efficiency of WLAN networks. Currently in development, this project has the goal of providing 4x the throughput of 802.11ac.

802.11ay

IEEE 802.11ay is a standard that is being developed. It is an amendment that defines a new physical layer for 802.11 networks to operate in the 60 GHz millimeter wave spectrum. It will be an extension of the existing 11ad, aimed to extend the throughput, range and use-cases. The main use-cases include: indoor operation, out-door back-haul and short range communications. The peak transmission rate of 802.11ay is 20 Gbit/s. The main extensions include: channel bonding (2, 3 and 4), MIMO and higher modulation schemes.

802.11-2016

IEEE 802.11-2016 is a revision based on IEEE 802.11-2012, incorporating 5 amendments (11ae, 11aa, 11ad, 11ac, 11af). In addition, existing MAC and PHY functions have been enhanced and obsolete features were removed or marked for removal. Some clauses and annexes have been renumbered.

Common Misunderstandings About Achievable Throughput

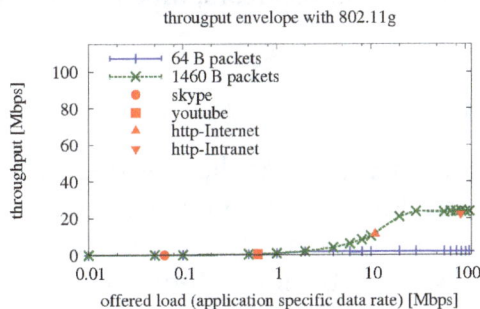

Graphical representation of Wi-Fi application specific (UDP) performance envelope 2.4 GHz band, with 802.11g

Across all variations of 802.11, maximum achievable throughputs are given either based on measurements under ideal conditions or in the layer-2 data rates. This, however, does not apply to typical deployments in which data is being transferred between two endpoints, of which at least one is typically connected to a wired infrastructure and the other endpoint is connected to an infrastructure via a wireless link.

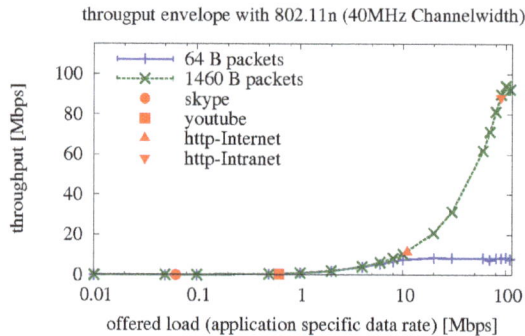

Graphical representation of Wi-Fi application specific (UDP) performance envelope 2.4 GHz band, with 802.11n with 40MHz.

This means that, typically, data frames pass an 802.11 (WLAN) medium, and are being converted to 802.3 (Ethernet) or vice versa. Due to the difference in the frame (header) lengths of these two media, the application's packet size determines the speed of the data transfer. This means applications that use small packets (e.g., VoIP) create dataflows with high-overhead traffic (i.e., a low goodput). Other factors that contribute to the overall application data rate are the speed with which the application transmits the packets (i.e., the data rate) and, of course, the energy with which the wireless signal is received. The latter is determined by distance and by the configured output power of the communicating devices.

The same references apply to the attached graphs that show measurements of UDP throughput. Each represents an average (UDP) throughput (please note that the error bars are there, but barely visible due to the small variation) of 25 measurements. Each is with a specific packet size (small or large) and with a specific data rate (10 kbit/s – 100 Mbit/s). Markers for traffic profiles of common applications are included as well. These figures assume there are no packet errors, which if occurring will lower transmission rate further.

Channels and Frequencies

802.11b, 802.11g, and 802.11n-2.4 utilize the 2.400–2.500 GHz spectrum, one of the ISM bands. 802.11a and 802.11n use the more heavily regulated 4.915–5.825 GHz band. These are commonly referred to as the "2.4 GHz and 5 GHz bands" in most sales literature. Each spectrum is sub-divided into *channels* with a center frequency and bandwidth, analogous to the way radio and TV broadcast bands are sub-divided.

The 2.4 GHz band is divided into 14 channels spaced 5 MHz apart, beginning with channel 1, which is centered on 2.412 GHz. The latter channels have additional restrictions or are unavailable for use in some regulatory domains.

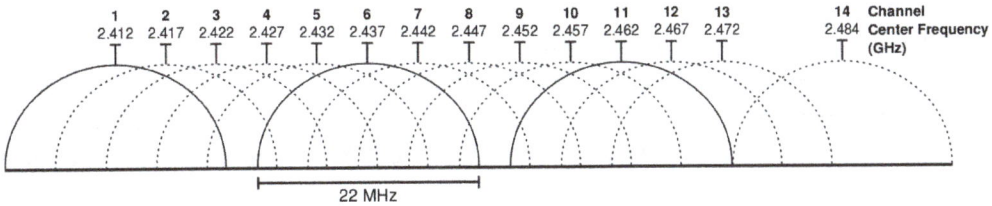

Graphical representation of Wi-Fi channels in the 2.4 GHz band

The channel numbering of the 5.725–5.875 GHz spectrum is less intuitive due to the differences in regulations between countries.

Channel Spacing within the 2.4 GHz Band

In addition to specifying the channel center frequency, 802.11 also specifies (in Clause 17) a spectral mask defining the permitted power distribution across each channel. The mask requires the signal be attenuated a minimum of 20 dB from its peak amplitude at ±11 MHz from the centre frequency, the point at which a channel is effectively 22 MHz wide. One consequence is that stations can use only every fourth or fifth channel without overlap.

Availability of channels is regulated by country, constrained in part by how each country allocates radio spectrum to various services. At one extreme, Japan permits the use of all 14 channels for 802.11b, and 1–13 for 802.11g/n-2.4. Other countries such as Spain initially allowed only channels 10 and 11, and France allowed only 10, 11, 12, and 13; however, they now allow channels 1 through 13. North America and some Central and South American countries allow only 1 through 11.

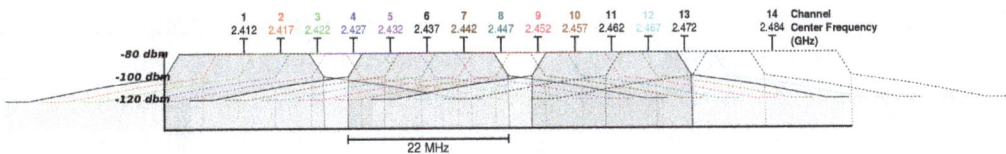

Spectral masks for 802.11g channels 1–14 in the 2.4 GHz band

Since the spectral mask defines only power output restrictions up to ±11 MHz from the center frequency to be attenuated by −50 dBr, it is often assumed that the energy of the channel extends no further than these limits. It is more correct to say that, given the separation between channels, the overlapping signal on any channel should be sufficiently attenuated to minimally interfere with a transmitter on any other channel. Due to the near-far problem a transmitter can impact (desense) a receiver on a "non-overlapping" channel, but only if it is close to the victim receiver (within a meter) or operating above allowed power levels.

Confusion often arises over the amount of channel separation required between transmitting devices. 802.11b was based on DSSS modulation and utilized a channel bandwidth of 22 MHz, resulting in *three* "non-overlapping" channels (1, 6, and 11). 802.11g was based on OFDM modulation and utilized a channel bandwidth of 20 MHz. This occasionally leads to the belief that *four* "non-overlapping" channels (1, 5, 9, and 13) exist under 802.11g, although this is not the case as per 17.4.6.3 Channel Numbering of operating channels of the IEEE Std 802.11 (2012), which states "In a multiple cell network topology, overlapping and/or adjacent cells using different channels can operate simultaneously without interference if the distance between the center frequencies is at least 25 MHz."

This does not mean that the technical overlap of the channels recommends the nonuse of overlapping channels. The amount of interference seen on a configuration using channels 1, 5, 9, and 13 can have very small difference from a three-channel configuration, and in the paper entitled "Effect of adjacent-channel interference in IEEE 802.11 WLANs" by Villegas this is also demonstrated.

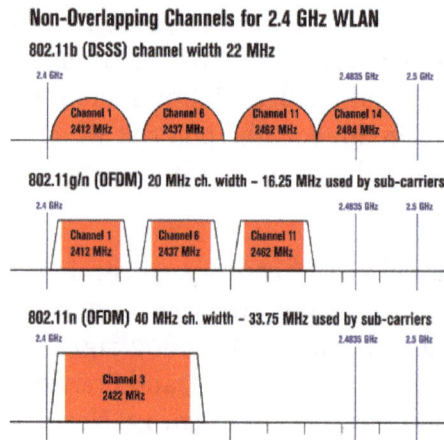

802.11 non-overlapping channels for 2.4GHz. Covers 802.11b,g,n

Although the statement that channels 1, 5, 9, and 13 are "non-overlapping" is limited to spacing or product density, the concept has some merit in limited circumstances. Special care must be taken to adequately space AP cells, since overlap between the channels may cause unacceptable degradation of signal quality and throughput. If more advanced equipment such as spectral analyzers are available, overlapping channels may be used under certain circumstances. This way, more channels are available.

Regulatory Domains and Legal Compliance

IEEE uses the phrase *regdomain* to refer to a legal regulatory region. Different countries define different levels of allowable transmitter power, time that a channel can be occupied, and different available channels. Domain codes are specified for the United States, Canada, ETSI (Europe), Spain, France, Japan, and China.

Most Wi-Fi certified devices default to *regdomain* 0, which means least common denominator settings, i.e., the device will not transmit at a power above the allowable power in any nation, nor will it use frequencies that are not permitted in any nation.

The *regdomain* setting is often made difficult or impossible to change so that the end users do not conflict with local regulatory agencies such as the United States' Federal Communications Commission.

Layer 2 – Datagrams

The datagrams are called *frames*. Current 802.11 standards specify frame types for use in transmission of data as well as management and control of wireless links.

Frames are divided into very specific and standardized sections. Each frame consists of a MAC header, payload, and frame check sequence (FCS). Some frames may not have a payload.

The first two bytes of the MAC header form a frame control field specifying the form and function of the frame. This frame control field is subdivided into the following subfields:

- Protocol Version: Two bits representing the protocol version. Currently used protocol version is zero. Other values are reserved for future use.

- Type: Two bits identifying the type of WLAN frame. Control, Data, and Management are various frame types defined in IEEE 802.11.

- Subtype: Four bits providing additional discrimination between frames. Type and Subtype are used together to identify the exact frame.

- ToDS and FromDS: Each is one bit in size. They indicate whether a data frame is headed for a distribution system. Control and management frames set these values to zero. All the data frames will have one of these bits set. However communication within an Independent Basic Service Set (IBSS) network always set these bits to zero.

- More Fragments: The More Fragments bit is set when a packet is divided into multiple frames for transmission. Every frame except the last frame of a packet will have this bit set.

- Retry: Sometimes frames require retransmission, and for this there is a Retry bit that is set to one when a frame is resent. This aids in the elimination of duplicate frames.

- Power Management: This bit indicates the power management state of the sender after the completion of a frame exchange. Access points are required to manage the connection, and will never set the power-saver bit.

- More Data: The More Data bit is used to buffer frames received in a distributed system. The access point uses this bit to facilitate stations in power-saver mode. It indicates that at least one frame is available, and addresses all stations connected.

- Protected Frame: The Protected Frame bit is set to one if the frame body is encrypted by a protection mechanism such as Wired Equivalent Privacy (WEP), Wi-Fi Protected Access (WPA), or Wi-FI Protected Access II (WPA2).

- Order: This bit is set only when the "strict ordering" delivery method is employed. Frames and fragments are not always sent in order as it causes a transmission performance penalty.

The next two bytes are reserved for the Duration ID field. This field can take one of three forms: Duration, Contention-Free Period (CFP), and Association ID (AID).

An 802.11 frame can have up to four address fields. Each field can carry a MAC address. Address 1 is the receiver, Address 2 is the transmitter, Address 3 is used for filtering purposes by the receiver.

The remaining fields of the header are:

- The Sequence Control field is a two-byte section used for identifying message order as well as eliminating duplicate frames. The first 4 bits are used for the fragmentation number, and the last 12 bits are the sequence number.

- An optional two-byte Quality of Service control field that was added with 802.11e.

The payload or frame body field is variable in size, from 0 to 2304 bytes plus any overhead from security encapsulation, and contains information from higher layers.

The Frame Check Sequence (FCS) is the last four bytes in the standard 802.11 frame. Often referred to as the Cyclic Redundancy Check (CRC), it allows for integrity check of retrieved frames. As frames are about to be sent, the FCS is calculated and appended. When a station receives a frame, it can calculate the FCS of the frame and compare it to the one received. If they match, it is assumed that the frame was not distorted during transmission.

Management Frames

Management frames allow for the maintenance of communication. Some common 802.11 subtypes include:

- Authentication frame: 802.11 authentication begins with the Wireless network interface card (WNIC) sending an authentication frame to the access point containing its identity. With an open system authentication, the WNIC sends only a single authentication frame, and the access point responds with an authentication frame of its own indicating acceptance or rejection. With shared key authentication, after the WNIC sends its initial authentication request it will receive an authentication frame from the access point containing challenge text. The WNIC sends an authentication frame containing the encrypted version of the challenge text to the access point. The access point ensures the text was encrypted with the correct key by decrypting it with its own key. The result of this process determines the WNIC's authentication status.

- Association request frame: Sent from a station it enables the access point to allocate resources and synchronize. The frame carries information about the WNIC, including supported data rates and the SSID of the network the station wishes to associate with. If the request is accepted, the access point reserves memory and establishes an association ID for the WNIC.

- Association response frame: Sent from an access point to a station containing the acceptance or rejection to an association request. If it is an acceptance, the frame will contain information such an association ID and supported data rates.

- Beacon frame: Sent periodically from an access point to announce its presence and provide the SSID, and other parameters for WNICs within range.

- Deauthentication frame: Sent from a station wishing to terminate connection from another station.

- Disassociation frame: Sent from a station wishing to terminate connection. It's an elegant way to allow the access point to relinquish memory allocation and remove the WNIC from the association table.

- Probe request frame: Sent from a station when it requires information from another station.

- Probe response frame: Sent from an access point containing capability information, supported data rates, etc., after receiving a probe request frame.

- Reassociation request frame: A WNIC sends a reassociation request when it drops from range of the currently associated access point and finds another access point with a stronger signal. The new access point coordinates the forwarding of any information that may still be contained in the buffer of the previous access point.

- Reassociation response frame: Sent from an access point containing the acceptance or rejection to a WNIC reassociation request frame. The frame includes information required for association such as the association ID and supported data rates.

Information Elements

2. In terms of ICT, an Information Element (IE) is a part of management frames in the IEEE 802.11 wireless LAN protocol. IEs are a device's way to transfer descriptive information about itself inside management frames. There are usually several IEs inside each such frame, and each is built of TLVs mostly defined outside the basic IEEE 802.11 specification.

The common structure of an IE is as follows:

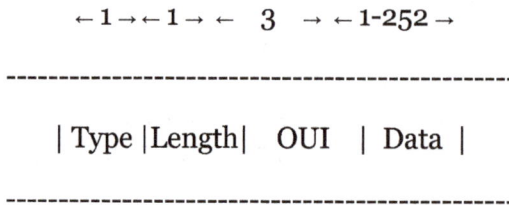

$$\leftarrow 1 \rightarrow \leftarrow 1 \rightarrow \leftarrow \quad 3 \quad \rightarrow \leftarrow 1\text{-}252 \rightarrow$$

```
-----------------------------------------------

| Type |Length|   OUI   | Data |

-----------------------------------------------
```

Whereas the OUI (organizationally unique identifier) is used only when necessary to the protocol being used, and the data field holds the TLVs relevant to that IE.

Control Frames

Control frames facilitate in the exchange of data frames between stations. Some common 802.11 control frames include:

- Acknowledgement (ACK) frame: After receiving a data frame, the receiving station will send an ACK frame to the sending station if no errors are found. If the sending station doesn't receive an ACK frame within a predetermined period of time, the sending station will resend the frame.

- Request to Send (RTS) frame: The RTS and CTS frames provide an optional collision reduction scheme for access points with hidden stations. A station sends a RTS frame as the first step in a two-way handshake required before sending data frames.

- Clear to Send (CTS) frame: A station responds to an RTS frame with a CTS frame. It provides clearance for the requesting station to send a data frame. The CTS provides collision control management by including a time value for which all other stations are to hold off transmission while the requesting station transmits.

Data Frames

Data frames carry packets from web pages, files, etc. within the body. The body begins with an IEEE 802.2 header, with the Destination Service Access Point (DSAP) specifying the protocol; however, if the DSAP is hex AA, the 802.2 header is followed by a Subnetwork Access Protocol (SNAP) header, with the Organizationally Unique Identifier (OUI) and protocol ID (PID) fields specifying the protocol. If the OUI is all zeroes, the protocol ID field is an EtherType value. Almost all 802.11 data frames use 802.2 and SNAP headers, and most use an OUI of 00:00:00 and an EtherType value.

Similar to TCP congestion control on the internet, frame loss is built into the operation of 802.11. To select the correct transmission speed or Modulation and Coding Scheme, a rate control algorithm may test different speeds. The actual packet loss rate of an Access points vary widely for different link conditions. There are variations in the loss rate experienced on production Access points, between 10% and 80%, with 30% being a common average. It is important to be aware that the link layer should recover these lost frames. If the sender does not receive an Acknowledgement (ACK) frame, then it will be resent.

Standards and Amendments

Within the IEEE 802.11 Working Group, the following IEEE Standards Association Standard and Amendments exist:

- IEEE 802.11-1997: The WLAN standard was originally 1 Mbit/s and 2 Mbit/s, 2.4 GHz RF and infrared (IR) standard (1997), all the others listed below are Amendments to this standard, except for Recommended Practices 802.11F and 802.11T.

- IEEE 802.11a: 54 Mbit/s, 5 GHz standard (1999, shipping products in 2001)

- IEEE 802.11b: Enhancements to 802.11 to support 5.5 Mbit/s and 11 Mbit/s (1999)

- IEEE 802.11c: Bridge operation procedures; included in the IEEE 802.1D standard (2001)

- IEEE 802.11d: International (country-to-country) roaming extensions (2001)

- IEEE 802.11e: Enhancements: QoS, including packet bursting (2005)

- IEEE 802.11F: Inter-Access Point Protocol (2003) Withdrawn February 2006

- IEEE 802.11g: 54 Mbit/s, 2.4 GHz standard (backwards compatible with b) (2003)

- IEEE 802.11h: Spectrum Managed 802.11a (5 GHz) for European compatibility (2004)

- IEEE 802.11i: Enhanced security (2004)

- IEEE 802.11j: Extensions for Japan (2004)

- IEEE 802.11-2007: A new release of the standard that includes amendments a, b, d, e, g, h, i, and j. (July 2007)

- IEEE 802.11k: Radio resource measurement enhancements (2008)

- IEEE 802.11n: Higher-throughput improvements using MIMO (multiple-input, multiple-output antennas) (September 2009)

- IEEE 802.11p: WAVE—Wireless Access for the Vehicular Environment (such as ambulances and passenger cars) (July 2010)

- IEEE 802.11r: Fast BSS transition (FT) (2008)

- IEEE 802.11s: Mesh Networking, Extended Service Set (ESS) (July 2011)

- IEEE 802.11T: Wireless Performance Prediction (WPP)—test methods and metrics Recommendation cancelled

- IEEE 802.11u: Improvements related to HotSpots and 3rd-party authorization of clients, e.g., cellular network offload (February 2011)

- IEEE 802.11v: Wireless network management (February 2011)

- IEEE 802.11w: Protected Management Frames (September 2009)

- IEEE 802.11y: 3650–3700 MHz Operation in the U.S. (2008)

- IEEE 802.11z: Extensions to Direct Link Setup (DLS) (September 2010)

- IEEE 802.11-2012: A new release of the standard that includes amendments k, n, p, r, s, u, v, w, y, and z (March 2012)

- IEEE 802.11aa: Robust streaming of Audio Video Transport Streams (June 2012)

- IEEE 802.11ac: Very High Throughput <6 GHz; potential improvements over 802.11n: better modulation scheme (expected ~10% throughput increase), wider channels (estimate in future time 80 to 160 MHz), multi user MIMO; (December 2013)

- IEEE 802.11ad: Very High Throughput 60 GHz (December 2012)

- IEEE 802.11ae: Prioritization of Management Frames (March 2012)

- IEEE 802.11af: TV Whitespace (February 2014)

- IEEE 802.11ah: Sub-1 GHz license exempt operation (e.g., sensor network, smart metering) (December 2016)

- IEEE 802.11ai: Fast Initial Link Setup (December 2016)

- IEEE 802.11-2016: A new release of the standard that includes amendments ae, aa, ad, ac, and af (December 2016)

In Process

- IEEE 802.11aj: China Millimeter Wave *(~ December 2017)*

- IEEE 802.11ak: General Links *(~ December 2017)*

- IEEE 802.11aq: Pre-association Discovery *(~ August 2017)*

- IEEE 802.11ax: High Efficiency WLAN *(~ July 2019)*

- IEEE 802.11ay: Enhancements for Ultra High Throughput in and around the 60 GHz Band *(~ November 2019)*

- IEEE 802.11az: Next Generation Positioning *(~ March 2021)*

- IEEE 802.11ba: Wake Up Radio *(~ July 2020)*

802.11F and 802.11T are recommended practices rather than standards, and are capitalized as such.

802.11m is used for standard maintenance. 802.11ma was completed for 802.11-2007, 802.11mb for 802.11-2012, and 802.11mc for 802.11-2016.

Standard Vs. Amendment

Both the terms "standard" and "amendment" are used when referring to the different variants of IEEE standards.

As far as the IEEE Standards Association is concerned, there is only one current standard; it is denoted by IEEE 802.11 followed by the date that it was published. IEEE 802.11-2016 is the only version currently in publication, superseding previous releases. The standard is updated by means of amendments. Amendments are created by task groups (TG). Both the task group and their finished document are denoted by 802.11 followed by a non-capitalized letter, for example, IEEE 802.11a and IEEE 802.11b. Updating 802.11 is the responsibility of task group m. In order to create a new version, TGm combines the previous version of the standard and all published amendments. TGm also provides clarification and interpretation to industry on published documents. New versions of the IEEE 802.11 were published in 1999, 2007, 2012, and 2016.

Nomenclature

Various terms in 802.11 are used to specify aspects of wireless local-area networking operation, and may be unfamiliar to some readers.

For example, Time Unit (usually abbreviated TU) is used to indicate a unit of time equal to 1024 microseconds. Numerous time constants are defined in terms of TU (rather than the nearly equal millisecond).

Also the term "Portal" is used to describe an entity that is similar to an 802.1H bridge. A Portal provides access to the WLAN by non-802.11 LAN STAs.

Community Networks

With the proliferation of cable modems and DSL, there is an ever-increasing market of people who wish to establish small networks in their homes to share their broadband Internet connection.

Many hotspot or free networks frequently allow anyone within range, including passersby outside, to connect to the Internet. There are also efforts by volunteer groups to establish wireless community networks to provide free wireless connectivity to the public.

Security

In 2001, a group from the University of California, Berkeley presented a paper describing weaknesses in the 802.11 Wired Equivalent Privacy (WEP) security mechanism defined in the original standard; they were followed by Fluhrer, Mantin, and Shamir's paper titled "Weaknesses in the Key Scheduling Algorithm of RC4". Not long after, Adam Stubblefield and AT&T publicly announced the first verification of the attack. In the attack, they were able to intercept transmissions and gain unauthorized access to wireless networks.

The IEEE set up a dedicated task group to create a replacement security solution, 802.11i (previously this work was handled as part of a broader 802.11e effort to enhance the MAC layer). The Wi-Fi Alliance announced an interim specification called Wi-Fi Protected Access (WPA) based on a subset of the then current IEEE 802.11i draft. These started to appear in products in mid-2003. IEEE 802.11i (also known as WPA2) itself was ratified in June 2004, and uses the Advanced Encryption Standard AES, instead of RC4, which was used in WEP. The modern recommended encryption for the home/consumer space is WPA2 (AES Pre-Shared Key), and for the enterprise space is WPA2 along with a RADIUS authentication server (or another type of authentication server) and a strong authentication method such as EAP-TLS.

In January 2005, the IEEE set up yet another task group "w" to protect management and broadcast frames, which previously were sent unsecured. Its standard was published in 2009.

In December 2011, a security flaw was revealed that affects some wireless routers with a specific implementation of the optional Wi-Fi Protected Setup (WPS) feature. While WPS is not a part of 802.11, the flaw allows an attacker within the range of the wireless router to recover the WPS PIN and, with it, the router's 802.11i password in a few hours.

In late 2014, Apple announced that its iOS 8 mobile operating system would scramble MAC addresses during the pre-association stage to thwart retail footfall tracking made possible by the regular transmission of uniquely identifiable probe requests.

Non-standard 802.11 Extensions and Equipment

Many companies implement wireless networking equipment with non-IEEE standard 802.11 extensions either by implementing proprietary or draft features. These changes may lead to incompatibilities between these extensions.

Wi-Fi Protected Setup

The WPS push button (center, blue) on a wireless router showing the symbol defined by the Wi-Fi Alliance for this function. This statement works for WPS users.

Wi-Fi Protected Setup (WPS; originally, Wi-Fi Simple Config) is a network security standard to create a secure wireless home network.

Created by the Wi-Fi Alliance and introduced in 2006, the goal of the protocol is to allow home users who know little of wireless security and may be intimidated by the available security options to set up Wi-Fi Protected Access, as well as making it easy to add new devices to an existing network without entering long passphrases. Prior to the standard, several competing solutions were developed by different vendors to address the same need.

A major security flaw was revealed in December 2011 that affects wireless routers with the WPS PIN feature, which most recent models have enabled by default. The flaw allows a remote attacker to recover the WPS PIN in a few hours with a brute-force attack and, with the WPS PIN, the network's WPA/WPA2 pre-shared key. Users have been urged to turn off the WPS PIN feature, although this may not be possible on some router models.

Modes

The standard emphasizes usability and security, and allows four modes in a home network for adding a new device to the network.

PIN Method

In which a PIN has to be read from either a sticker or display on the new wireless device. This PIN must then be entered at the "representant" of the network, usually the network's access point. Alt access point may be entered into the new device. This method is the mandatory baseline mode and everything must support it. The Wi-Fi Direct specification supersedes this requirement by stating that all devices with a keypad or display must support the PIN method.

Push Button Method

In which the user has to push a button, either an actual or virtual one, on both the access point and the new wireless client device. On most devices, this discovery mode turns itself off as soon as a connection is established or after a delay (typically 2 minutes or less), whichever comes first, thereby minimizing its vulnerability. Support of this mode is mandatory for access points and optional for connecting devices. The Wi-Fi Direct specification supersedes this requirement by stating that all devices must support the push button method.

Near-field Communication Method

In which the user has to bring the new client close to the access point to allow a near field communication between the devices. NFC Forum–compliant RFID tags can also be used. Support of this mode is optional.

USB Method

In which the user uses a USB flash drive to transfer data between the new client device and the network's access point. Support of this mode is optional, but deprecated.

The last two modes are usually referred to as out-of-band methods as there is a transfer of information by a channel other than the Wi-Fi channel itself. Only the first two modes are currently covered by the WPS certification. The USB method has been deprecated and is not part of the Alliance's certification testing.

Some wireless access points have a dual-function WPS button, and holding this button down for a shorter or longer time may have other functions, such as factory-reset or toggling WiFi.

Some manufacturers, such as Netgear, use a different logo and/or name for Wi-Fi Protected Setup; the Wi-Fi Alliance recommends the use of the Wi-Fi Protected Setup Identifier Mark on the hardware button for this function.

Technical Architecture

The WPS protocol defines three types of devices in a network:

Registrar

A device with the authority to issue and revoke access to a network; it may be integrated into a wireless access point (AP), or provided as a separate device.

Enrollee

A client device seeking to join a wireless network.

AP

An access point functioning as a proxy between a registrar and an enrollee.

The WPS standard defines three basic scenarios that involve components listed above:

AP with integrated registrar capabilities configures an Enrollee Station (STA)

In this case, the session will run on the wireless medium as a series of EAP request/response messages, ending with the AP disassociating from the STA and waiting for the STA to reconnect with its new configuration (handed to it by the AP just before).

Registrar STA Configures the AP as an Enrollee

This case is subdivided in two aspects: first, the session could occur on either a wired or wireless medium, and second, the AP could already be configured by the time the registrar found it. In the case of a wired connection between the devices, the protocol runs over Universal Plug and Play (UPnP), and both devices will have to support UPnP for that purpose. When running over UPnP, a shortened version of the protocol is run (only two messages) as no authentication is required other than that of the joined wired medium. In the case of a wireless medium, the session of the protocol is very similar to the internal registrar scenario, but with opposite roles. As to the configuration state of the AP, the registrar is expected to ask the user whether to reconfigure the AP or keep its current settings, and can decide to reconfigure it even if the AP describes itself as configured. Multiple registrars should have the ability to connect to the AP. UPnP is intended to apply only to a wired medium, while actually it applies to any interface to which an IP connection can be set up. Thus, having manually set up a wireless connection, the UPnP can be used over it in the same manner as with the wired connection.

Registrar STA Configures Enrollee STA

In this case the AP stands in the middle and acts as an authenticator, meaning it only proxies the relevant messages from side to side.

Protocol

The WPS protocol consists of a series of EAP message exchanges that are triggered by a user action, relying on an exchange of descriptive information that should precede that user's action. The descriptive information is transferred through a new Information Element (IE) that is added to the beacon, probe response, and optionally to the probe request and association request/response messages. Other than purely informative type-length-values, those IEs will also hold the possible and the currently deployed configuration methods of the device.

After this communication of the device capabilities from both ends, the user initiates the actual protocol session. The session consists of eight messages that are followed, in the case of a successful session, by a message to indicate that the protocol is completed. The exact stream of messages may change when configuring different kinds of devices (AP or STA), or when using different physical media (wired or wireless).

Band or Radio Selection

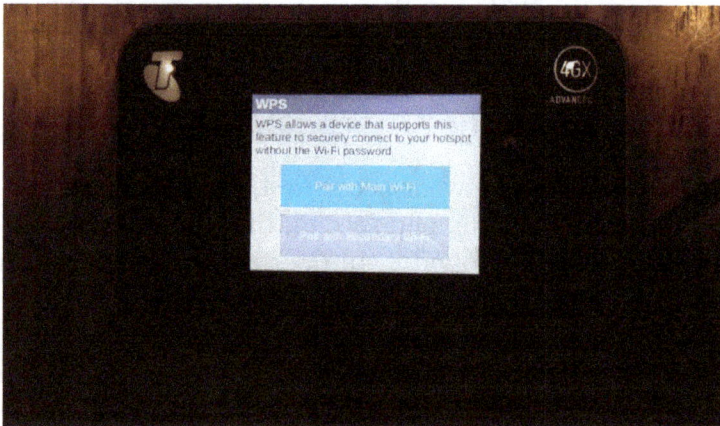

A Telstra 4GX Advanced III mobile broadband device showing WPS pairing options for a particular radio/band.

Some devices with dual-band wireless network connectivity do not allow the user to select the 2.4 GHz or 5 GHz band (or even a particular radio or SSID) when using Wi-Fi Protected Setup, unless the wireless access point has separate WPS button for each band or radio; however, a number of later wireless routers with multiple frequency bands and/or radios allow the establishment of a WPS session for a specific band and/or radio for connection with clients which cannot have the SSID or band (e.g., 2.4/5 GHz) explicitly selected by the user on the client for connection with WPS (e.g. pushing the 5 GHz, where supported, WPS button on the wireless router will force a client device to connect via WPS on only the 5 GHz band after a WPS session has been established by the client device which cannot explicitly allow the selection of wireless network and/or band for the WPS connection method).

Vulnerabilities

Online Brute-force Attack

In December 2011, researcher Stefan Viehböck reported a design and implementation flaw that makes brute-force attacks against PIN-based WPS feasible to be performed on WPS-enabled Wi-Fi networks. A successful attack on WPS allows unauthorized parties to gain access to the network, and the only effective workaround is to disable WPS. The vulnerability centers around the acknowledgement messages sent between the registrar and enrollee when attempting to validate a PIN, which is an eight-digit number used to add new WPA enrollees to the network. Since the last digit is a checksum of the previous digits, there are seven unknown digits in each PIN, yielding $10^7 = 10,000,000$ possible combinations.

When an enrollee attempts to gain access using a PIN, the registrar reports the validity of the first and second halves of the PIN separately. Since the first half of the pin consists of four digits (10,000 possibilities) and the second half has only three active digits (1000 possibilities), at most 11,000 guesses are needed before the PIN is recovered. This is a reduction by three orders of magnitude from the number of PINs that would be required to be tested. As a result, an attack can be completed in under four hours. The ease or difficulty of exploiting this flaw is implementation-dependent, as Wi-Fi router manufacturers could defend against such attacks by slowing or disabling the WPS feature after several failed PIN validation attempts.

A young developer based out of a small town in eastern New Mexico created a tool that exploits this vulnerability to prove that the attack is feasible. The tool was then purchased by Tactical Network Solutions in Maryland for 1.5 million dollars. They state that they have known about the vulnerability since early 2011 and has been using it.

In some devices, disabling WPS in the user interface does not result in the feature actually being disabled, and the device remains vulnerable to this attack. Firmware updates have been released for some of these devices allowing WPS to be disabled completely. Vendors could also patch the vulnerability by adding a lock-down period if the Wi-Fi access point detects a brute-force attack in progress, which disables the PIN method for long enough to make the attack impractical.

Offline Brute-force Attack

In the summer of 2014, Dominique Bongard discovered what he called the *Pixie Dust* attack. This attack works only for the default WPS implementation of several wireless chip makers, including Ralink, MediaTek, Realtek and Broadcom. The attack focuses on a lack of randomization when generating the E-S1 and E-S2 "secret" nonces. Knowing these two nonces, the PIN can be recovered within a couple of minutes. A tool called *pixiewps* has been developed and a new version of Reaver has been developed to automate the process.

Since both the access point and client (enrollee and registrar, respectively) need to prove they know the PIN to make sure the client is not connecting to a rogue AP, the attacker already has two hashes that contain each half of the PIN, and all they need is to brute-force the actual PIN. The access point sends two hashes, E-Hash1 and E-Hash2, to the client, proving that it also knows the PIN. E-Hash1 and E-Hash2 are hashes of (E-S1 | PSK1 | PKe | PKr) and (E-S2 | PSK2 | PKe | PKr), respectively. The hashing function is HMAC-SHA-256 and uses the "authkey" that is the key used to hash the data.

Physical Security Issues

All WPS methods are vulnerable to usage by an unauthorized user if the wireless access point is not kept in a secure area. Many wireless access points have security information (if it is factory-secured) and the WPS PIN printed on them; this PIN is also often found in the configuration menus of the wireless access point. If this PIN cannot be changed or disabled, the only remedy is to get a firmware update to enable the PIN to be changed, or to replace the wireless access point.

It is possible to extract a wireless passphrase with the following methods using no special tools:

- A wireless passphrase can be extracted using WPS under Windows Vista and newer versions of Windows, under administrative privileges by connecting with this method then bringing up the properties for this wireless network and clicking on "show characters".

- A simple exploit in the Intel PROset wireless client utility can reveal the wireless passphrase when WPS is used, after a simple move of the dialog box which asks if you want to reconfigure this access point.

A generic service bulletin regarding physical security for wireless access points.

A wireless router showing printed preset security information including the Wi-Fi Protected Setup PIN.

And, since the WPS function allows connection with only a press of the WPS Button, if the wireless access point is not kept in a secure area, an unauthorized user can connect to WPS-enabled network, simply by a press of the WPS button to command the router to connect to any device requesting WPS connection, therefore make it possible to "export" network passwords.

References

- Marziah Karch (1 September 2010). Android for Work: Productivity for Professionals. Apress. ISBN 978-1-4302-3000-7. Retrieved 11 November 2012

- "Authorization of Spread Spectrum Systems Under Parts 15 and 90 of the FCC Rules and Regulations". Federal Communications Commission of the USA. June 18, 1985. Archived from the original (txt) on September 28, 2007. Retrieved 2007-08-31

- Wolter Lemstra; Vic Hayes; John Groenewegen (2010). The innovation journey of Wi-Fi: the road to global success. Cambridge University Press. p. 121. ISBN 978-0-521-19971-1. Retrieved October 6, 2011

- "Statement of Use, s/n 75799629, US Patent and Trademark Office Trademark Status and Document Retrieval". August 23, 2005. Retrieved 2014-09-21. first used the Certification Mark ... as early as August 1999

- Joshua Bardwell; Devin Akin (2005). Certified Wireless Network Administrator Official Study Guide (Third ed.). McGraw-Hill. p. 418. ISBN 978-0-07-225538-6

- Rubin, GJ; Das Munshi, J; Wessely, S (2005-03-01). "Electromagnetic Hypersensitivity: A Systematic Review of Provocation Studies". Psychosom Med. 67: 224–32. PMID 15784787. doi:10.1097/01.psy.0000155664.13300.64

- Sibthorpe, Clare (4 August 2016). "CSIRO Wi-Fi invention to feature in upcoming exhibition at National Museum of Australia". The Canberra Times. Retrieved 4 August 2016

- Ioanis Nikolaidis; Kui Wu (2010-07-13). Ad-Hoc, Mobile and Wireless Networks: 9th International Conference, ADHOC-NOW 2010, Edmonton, AB, Canada, August 20–22, 2010, Proceedings. Springer Science+Business Media. p. 202. ISBN 978-3-642-14784-5

- "Marvell supports new 'Wi-Fi Direct' standard across entire line of 802.11 devices". Press release. Marvell. October 14, 2009. Retrieved September 27, 2013

- Sergiu Nedevschi (2008). Maximizing performance in long distance wireless networks for developing regions. ProQuest. p. 28. ISBN 9781109096101

- W. David Gardner (2010-10-25). "Wi-Fi Direct Products Connect Without A Network". Informationweek.com. Retrieved 2013-07-30

- Pommer, hermann (2008-03-25). Roaming zwischen Wireless Local Area Networks. Saarbrücken: VDM Verlag. ISBN 978-3-8364-8708-5

- Moses, Asher (June 1, 2010). "CSIRO to reap 'lazy billion' from world's biggest tech companies". The Age. Melbourne. Retrieved 8 June 2010

Wireless Network: A Comprehensive Study

Wireless networks are computer networks which do not require cables. They are mainly administered by using radio communication. Wireless mesh network, free-space optical communication, municipal wireless network, wireless network interface controller, fixed wireless and wireless security are important topics related to the subject matter.

Wireless Network

Wireless icon

A wireless network is a computer network that uses wireless data connections between network nodes.

Wireless networking is a method by which homes, telecommunications networks and business installations avoid the costly process of introducing cables into a building, or as a connection between various equipment locations. Wireless telecommunications networks are generally implemented and administered using radio communication. This implementation takes place at the physical level (layer) of the OSI model network structure.

Examples of wireless networks include cell phone networks, wireless local area networks (WLANs), wireless sensor networks, satellite communication networks, and terrestrial microwave networks.

History

The first professional wireless network was developed under the brand ALOHAnet in 1969 at the University of Hawaii and became operational in June 1971. The first commercial wireless network was the WaveLAN product family, developed by NCR in 1986.

- 1991 2G cell phone network

- June 1997 802.11 "WiFi" protocol first release

- 1999 803.11 VoIP integration

Wireless Links

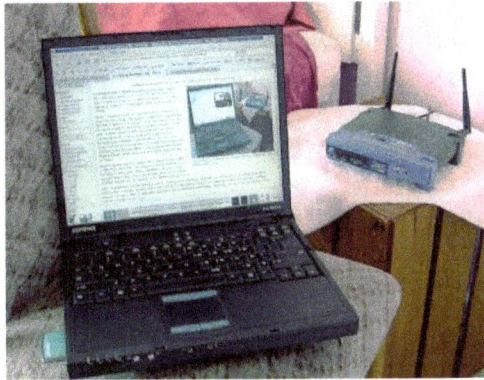

Computers are very often connected to networks using wireless links, e.g. WLANs

- *Terrestrial microwave* – Terrestrial microwave communication uses Earth-based transmitters and receivers resembling satellite dishes. Terrestrial microwaves are in the low gigahertz range, which limits all communications to line-of-sight. Relay stations are spaced approximately 48 km (30 mi) apart.

- *Communications satellites* – Satellites communicate via microwave radio waves, which are not deflected by the Earth's atmosphere. The satellites are stationed in space, typically in geosynchronous orbit 35,400 km (22,000 mi) above the equator. These Earth-orbiting systems are capable of receiving and relaying voice, data, and TV signals.

- *Cellular and PCS systems* use several radio communications technologies. The systems divide the region covered into multiple geographic areas. Each area has a low-power transmitter or radio relay antenna device to relay calls from one area to the next area.

- *Radio and spread spectrum technologies* – Wireless local area networks use a high-frequency radio technology similar to digital cellular and a low-frequency radio technology. Wireless LANs use spread spectrum technology to enable communication between multiple devices in a limited area. IEEE 802.11 defines a common flavor of open-standards wireless radio-wave technology known as Wifi.

- *Free-space optical communication* uses visible or invisible light for communications. In most cases, line-of-sight propagation is used, which limits the physical positioning of communicating devices.

Types of Wireless Networks

Wireless PAN

Wireless personal area networks (WPANs) internet devices within a relatively small area, that is generally within a person's reach. For example, both Bluetooth radio and invisible infrared light provides a WPAN for interconnecting a headset to a laptop. ZigBee also supports WPAN applications. Wi-Fi PANs are becoming commonplace (2010) as equipment designers start to integrate Wi-Fi into a variety of consumer electronic devices. Intel "My WiFi" and Windows 7 "virtual Wi-Fi" capabilities have made Wi-Fi PANs simpler and easier to set up and configure.

Wireless LAN

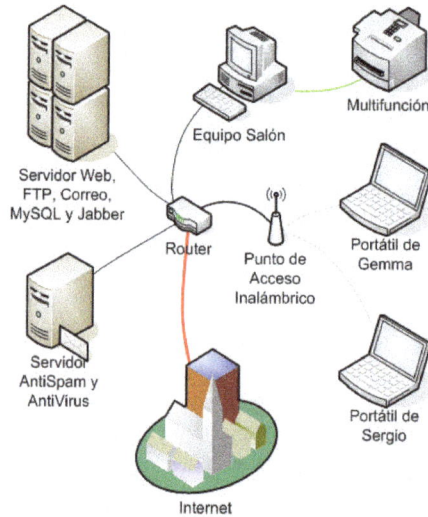

Wireless LANs are often used for connecting to local resources and to the Internet

A wireless local area network (WLAN) links two or more devices over a short distance using a wireless distribution method, usually providing a connection through an access point for internet access. The use of spread-spectrum or OFDM technologies may allow users to move around within a local coverage area, and still remain connected to the network.

Products using the IEEE 802.11 WLAN standards are marketed under the Wi-Fi brand name. Fixed wireless technology implements point-to-point links between computers or networks at two distant locations, often using dedicated microwave or modulated laser light beams over line of sight paths. It is often used in cities to connect networks in two or more buildings without installing a wired link.

Wireless ad Hoc Network

A wireless ad hoc network, also known as a wireless mesh network or mobile ad hoc network (MANET), is a wireless network made up of radio nodes organized in a mesh topology. Each node forwards messages on behalf of the other nodes and each node performs routing. Ad hoc networks can "self-heal", automatically re-routing around a node that has lost power. Various network layer protocols are needed to realize ad hoc mobile networks, such as Distance Sequenced Distance Vector routing, Associativity-Based Routing, Ad hoc on-demand Distance Vector routing, and Dynamic source routing.

Wireless MAN

Wireless metropolitan area networks are a type of wireless network that connects several wireless LANs.

- WiMAX is a type of Wireless MAN and is described by the IEEE 802.16 standard.

Wireless WAN

Wireless wide area networks are wireless networks that typically cover large areas, such as between neighbouring towns and cities, or city and suburb. These networks can be used to connect branch offices of business or as a public Internet access system. The wireless connections between access points are usually point to point microwave links using parabolic dishes on the 2.4 GHz band, rather than omnidirectional antennas used with smaller networks. A typical system contains base station gateways, access points and wireless bridging relays. Other configurations are mesh systems where each access point acts as a relay also. When combined with renewable energy systems such as photovoltaic solar panels or wind systems they can be stand alone systems.

Cellular Network

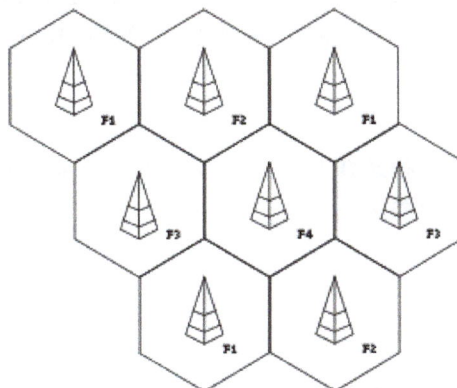

Example of frequency reuse factor or pattern 1/4

A cellular network or mobile network is a radio network distributed over land areas called cells, each served by at least one fixed-location transceiver, known as a cell site or base station. In a cellular network, each cell characteristically uses a different set of radio frequencies from all their immediate neighbouring cells to avoid any interference.

When joined together these cells provide radio coverage over a wide geographic area. This enables a large number of portable transceivers (e.g., mobile phones, pagers, etc.) to communicate with each other and with fixed transceivers and telephones anywhere in the network, via base stations, even if some of the transceivers are moving through more than one cell during transmission.

Although originally intended for cell phones, with the development of smartphones, cellular telephone networks routinely carry data in addition to telephone conversations:

- Global System for Mobile Communications (GSM): The GSM network is divided into three major systems: the switching system, the base station system, and the operation and support system. The cell phone connects to the base system station which then connects to the operation and support station; it then connects to the switching station where the call is transferred to where it needs to go. GSM is the most common standard and is used for a majority of cell phones.

- Personal Communications Service (PCS): PCS is a radio band that can be used by mobile phones in North America and South Asia. Sprint happened to be the first service to set up a PCS.

- D-AMPS: Digital Advanced Mobile Phone Service, an upgraded version of AMPS, is being phased out due to advancement in technology. The newer GSM networks are replacing the older system.

Global Area Network

A global area network (GAN) is a network used for supporting mobile across an arbitrary number of wireless LANs, satellite coverage areas, etc. The key challenge in mobile communications is handing off user communications from one local coverage area to the next. In IEEE Project 802, this involves a succession of terrestrial wireless LANs.

Space Network

Space networks are networks used for communication between spacecraft, usually in the vicinity of the Earth. The example of this is NASA's Space Network.

Different Uses

Some examples of usage include cellular phones which are part of everyday wireless networks, allowing easy personal communications. Another example, Intercontinental network systems, use radio satellites to communicate across the world. Emergency services such as the police utilize wireless networks to communicate effectively as well. Individuals and businesses use wireless networks to send and share data rapidly, whether it be in a small office building or across the world.

Properties

General

In a general sense, wireless networks offer a vast variety of uses by both business and home users.

"Now, the industry accepts a handful of different wireless technologies. Each wireless technology is defined by a standard that describes unique functions at both the Physical and the Data Link layers of the OSI model. These standards differ in their specified signaling methods, geographic ranges, and frequency usages, among other things. Such differences can make certain technologies better suited to home networks and others better suited to network larger organizations."

Performance

Each standard varies in geographical range, thus making one standard more ideal than the next depending on what it is one is trying to accomplish with a wireless network. The performance of wireless networks satisfies a variety of applications such as voice and video. The use of this technology also gives room for expansions, such as from 2G to 3G and, most recently, 4G technology, which stands for the fourth generation of cell phone mobile communications standards. As wireless networking has become commonplace, sophistication increases through configuration of network hardware and software, and greater capacity to send and receive larger amounts of data, faster, is achieved.

Space

Space is another characteristic of wireless networking. Wireless networks offer many advantages when it comes to difficult-to-wire areas trying to communicate such as across a street or river, a warehouse on the other side of the premises or buildings that are physically separated but operate as one. Wireless networks allow for users to designate a certain space which the network will be able to communicate with other devices through that network.

Space is also created in homes as a result of eliminating clutters of wiring. This technology allows for an alternative to installing physical network mediums such as TPs, coaxes, or fiber-optics, which can also be expensive.

Home

For homeowners, wireless technology is an effective option compared to Ethernet for sharing printers, scanners, and high-speed Internet connections. WLANs help save the cost of installation of cable mediums, save time from physical installation, and also creates mobility for devices connected to the network. Wireless networks are simple and require as few as one single wireless access point connected directly to the Internet via a router.

Wireless Network Elements

The telecommunication network at the physical layer also consists of many interconnected wireline network elements (NEs). These NEs can be stand-alone systems or products that are either supplied by a single manufacturer or are assembled by the service provider (user) or system integrator with parts from several different manufacturers.

Wireless NEs are the products and devices used by a wireless carrier to provide support for the backhaul network as well as a mobile switching center (MSC).

Reliable wireless service depends on the network elements at the physical layer to be protected against all operational environments and applications.

What are especially important are the NEs that are located on the cell tower to the base station (BS) cabinet. The attachment hardware and the positioning of the antenna and associated closures and cables are required to have adequate strength, robustness, corrosion resistance, and resistance against wind, storms, icing, and other weather conditions. Requirements for individual components, such as hardware, cables, connectors, and closures, shall take into consideration the structure to which they are attached.

Difficulties

Interferences

Compared to wired systems, wireless networks are frequently subject to electromagnetic interference. This can be caused by other networks or other types of equipment that generate radio waves that are within, or close, to the radio bands used for communication. Interference can degrade the signal or cause the system to fail.

Absorption and Reflection

Some materials cause absorption of electromagnetic waves, preventing it from reaching the receiver, in other cases, particularly with metallic or conductive materials reflection occurs. This can cause dead zones where no reception is available. Aluminium foiled thermal isolation in modern homes can easily reduce indoor mobile signals by 10 dB frequently leading to complaints about the bad reception of long-distance rural cell signals.

Multipath Fading

In multipath fading two or more different routes taken by the signal, due to reflections, can cause the signal to cancel out at certain locations, and to be stronger in other places (upfade).

Hidden Node Problem

The hidden node problem occurs in some types of network when a node is visible from a wireless access point (AP), but not from other nodes communicating with that AP. This leads to difficulties in media access control.

Shared Resource Problem

The wireless spectrum is a limited resource and shared by all nodes in the range of its transmitters. Bandwidth allocation becomes complex with multiple participating users. Often users are not aware that advertised numbers (e.g., for IEEE 802.11 equipment or LTE networks) are not their capacity, but shared with all other users and thus the individual user rate is far lower. With increasing demand, the capacity crunch is more and more likely to happen. User-in-the-loop (UIL) may be an alternative solution to ever upgrading to newer technologies for over-provisioning.

Capacity

Channel

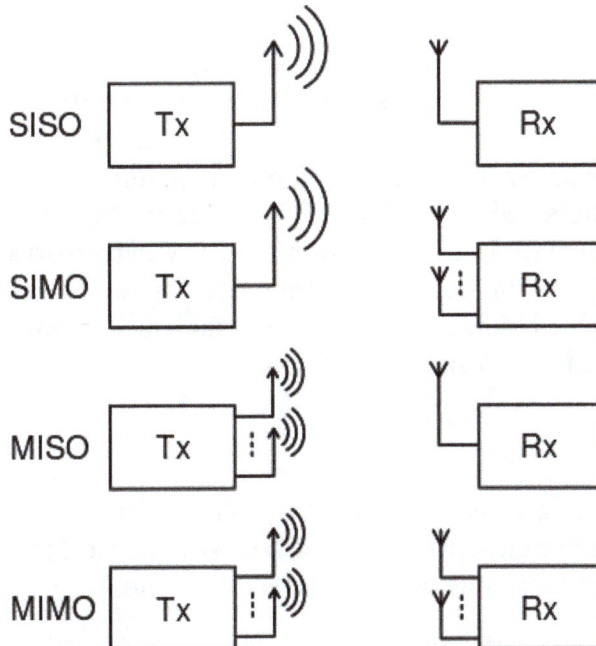

Understanding of SISO, SIMO, MISO and MIMO. Using multiple antennas and transmitting in different frequency channels can reduce fading, and can greatly increase the system capacity.

Shannon's theorem can describe the maximum data rate of any single wireless link, which relates to the bandwidth in hertz and to the noise on the channel.

One can greatly increase channel capacity by using MIMO techniques, where multiple aerials or multiple frequencies can exploit multiple paths to the receiver to achieve much higher throughput – by a factor of the product of the frequency and aerial diversity at each end.

Under Linux, the Central Regulatory Domain Agent (CRDA) controls the setting of channels.

Network

The total network bandwidth depends on how dispersive the medium is (more dispersive medium generally has better total bandwidth because it minimises interference), how many frequencies are available, how noisy those frequencies are, how many aerials are used and whether a directional antenna is in use, whether nodes employ power control and so on. there are two bands for now 2.4 ghz and 5 ghz. mostly 5 gigahertz band gives better connection and speed.

Cellular wireless networks generally have good capacity, due to their use of directional aerials, and their ability to reuse radio channels in non-adjacent cells. Additionally, cells can be made very small using low power transmitters this is used in cities to give network capacity that scales linearly with population density.

Safety

Wireless access points are also often close to humans, but the drop off in power over distance is fast, following the inverse-square law. The position of the United Kingdom's Health Protection Agency (HPA) is that "...radio frequency (RF) exposures from WiFi are likely to be lower than those from mobile phones." It also saw "...no reason why schools and others should not use WiFi equipment." In October 2007, the HPA launched a new "systematic" study into the effects of WiFi networks on behalf of the UK government, in order to calm fears that had appeared in the media in a recent period up to that time". Dr Michael Clark, of the HPA, says published research on mobile phones and masts does not add up to an indictment of WiFi.

Service Set (802.11 network)

In computer networking, a service set (SS) is a set consisting of all the devices associated with a consumer or enterprise IEEE 802.11 wireless local area network (WLAN). The service set can be local, independent, extended or mesh.

Service sets have an associated identifier, the service set identifier (SSID), which consists of 32 octets that frequently contains a human readable identifier of the network.

Basic Service Set (BSS)

The basic service set (BSS) provides the basic building-block of an 802.11 wireless LAN. In infrastructure mode, a single access point (AP) together with all associated stations (STAs) is called a BSS; different than the coverage of an access point, known as the basic service area (BSA). The access point acts as a master to control the stations within that BSS; the simplest BSS consists of one access point and one station.

The IEEE 802.11s amendment defined an additional protocol for wireless mesh networks. Only mesh STAs participate in mesh functionalities such as formation of the mesh BSS, path selection, and forwarding. Accordingly, a mesh STA is not a member of an *independent BSS* (*IBSS*) or of an *infrastructure BSS*. Consequently, mesh STAs do not communicate with non-mesh STAs. However, instead of existing independently, an MBSS can interconnect with other BSSs through the *distribution system* (*DS*). Mesh STAs can communicate with non-mesh STAs through a logical architectural component called a Mesh Gate.

With 802.11, one can alternatively set up an ad hoc network of client devices without a controlling access point; the result is called an independent BSS (IBSS).

Basic Service Set Identifier (BSSID)

Each BSS is uniquely identified by a basic service set identifier (BSSID).

For a BSS operating in infrastructure mode, the BSSID is the MAC address of the wireless access point (WAP) generated by combining the 24 bit Organization Unique Identifier (OUI, the manufacturer's identity) and the manufacturer's assigned 24-bit identifier for the radio chipset in the WAP. The BSSID is the formal name of the BSS and is always associated with only one BSS. Note, the MAC address concept is not limited to radio communication; wired networks use the very same 24+24 bit MAC address concept to uniquely identify the hosts.

The SSID is the informal (human) name of the BSS (just like a Windows Workgroup name). A BSS is functionally a contention domain as a local or workgroup network is functionally a broadcast domain.

A BSSID with a value of all 1s is used to indicate the wildcard BSSID, usable only during probe requests or for communications that take place outside the context of a BSS.

Extended Service Set (ESS)

An extended service set (ESS) is a set of two or more interconnected wireless BSSs that share the same SSID (network name), security credentials and integrated (that is, providing translation between 802.3 and 802.11 frames) wired local area networks that appear as a single BSS to the logical link control layer at any station associated with one of those BSSs. This facilitates mobile IP and fast secure roaming applications. Furthermore, the BSSs may work on the same channel, or work on different channels to boost aggregate throughput.

Extended Service Set Identifier (ESSID)

In an extended service set (ESS) each BSS still has its BSSID, however, the entire ESS uses only one SSID called an ESSID (to facilitate laptop and mobile internet device, MID, mobility; and voice over wifi, VoWiFi, roaming). For an IBSS, the BSSID is a locally administered MAC address generated from a 48-bit random number. The individual/group bit of the address is set to 0 (individual). The universal/local bit of the address is set to 1 (local).

Service Set Identifier (SSID)

Each BSS or ESS is identified by a service set identifier (SSID); a sequence of 0–32 octets. It is used as an identifier for a wireless LAN, and is intended to be unique for a particular area. Since this identifier must often be entered into devices manually by a human user, it is often a human-readable string and thus commonly called the "network name".

A common, albeit incorrect assumption, is that an SSID is a string of human-readable characters (such as ASCII), terminated by a NUL character (as in a C-string). SSIDs must be treated and handled as what they are, a sequence of 0–32 octets, some of which may not be human-readable. Note that the 2012 version of the 802.11 standard defines a primitive SSID encoding, an enumeration of UNSPECIFIED and UTF-8, indicating how the array of octets can be interpreted.

In an IBSS, the SSID is chosen by the client device that starts the network, and broadcasting of the SSID is performed in a pseudo-random order by all devices that are members of the network.

No Security of SSID Hiding

As a purported security enhancement, some access points allow a user to inhibit the broadcasting of their SSIDs, a tactic known as network cloaking; a station may then only join a BSS after the associated SSID has been specified explicitly. This tactic acts as a deterrent to the extent that it impedes casual wireless snooping, but is useless against widely available network scanners. Network cloaking is a form of security through obscurity and offers no protection against even the most basic attacks against a wireless network. Furthermore, the technique of SSID hiding forces wireless clients to probe for SSIDs everywhere they go, which makes them vulnerable to attackers who may set up a rogue access point emulating the probed network. As a consequence, hidden SSID probing at the client side involves greater security weaknesses than base station side beaconing.

Wireless WAN

A wireless wide area network (WWAN), is a form of wireless network. The larger size of a wide area network compared to a local area network requires differences in technology. Wireless networks of different sizes deliver data in the form of telephone calls, web pages, and streaming video.

A WWAN often differs from wireless local area network (WLAN) by using mobile telecommunication cellular network technologies such as LTE, WiMAX (often called a wireless metropolitan area network or WMAN), UMTS, CDMA2000, GSM, cellular digital packet data (CDPD) and Mobitex to transfer data. It can also use Local Multipoint Distribution Service (LMDS) or Wi-Fi to provide Internet access. These technologies are offered regionally, nationwide, or even globally and are provided by a wireless service provider. WWAN connectivity allows a user with a laptop and a WWAN card to surf the web, check email, or connect to a virtual private network (VPN) from anywhere within the regional boundaries of cellular service. Various computers can have integrated WWAN capabilities.

A WWAN may also be a closed network that covers a large geographic area. For example, a mesh network or MANET with nodes on building, tower, trucks, and planes could also be considered a WWAN.

A WWAN is also different from a low-power, low bit rate wireless WAN, (LPWAN), intended to carry small packets of information between things, often in the form of battery operated sensors.

Since radio communications systems do not provide a physically secure connection path, WWANs typically incorporate encryption and authentication methods to make them more secure. Unfortunately some of the early GSM encryption techniques were flawed, and security experts have issued warnings that cellular communication, including WWAN, is no longer secure. UMTS (3G) encryption was developed later and has yet to be broken.

Wireless Mesh Network

A wireless mesh network (WMN) is a communications network made up of radio nodes organized in a mesh topology. It is also a form of wireless ad hoc network.

A mesh refers to rich interconnection among devices or nodes. Wireless mesh networks often consist of mesh clients, mesh routers and gateways. Mobility of nodes is less frequent. If nodes were to constantly or frequently move, the mesh will spend more time updating routes than delivering data. In a wireless mesh network, topology tends to be more static, so that routes computation can converge and delivery of data to their destinations can occur. Hence, this is a low-mobility centralized form of wireless ad hoc network. Also, because it sometimes relies on static nodes to act as gateways, it is not a truly all-wireless ad hoc network.

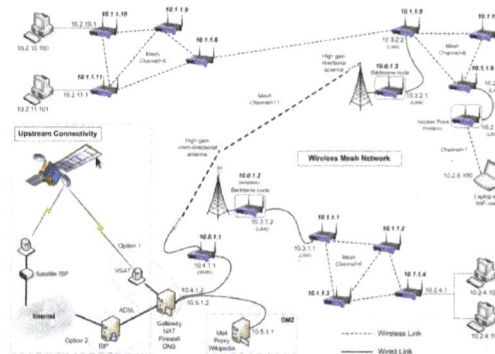

Diagram showing a possible configuration for a wired-wireless mesh network, connected upstream via a VSAT link.

The mesh clients are often laptops, cell phones and other wireless devices while the mesh routers forward traffic to and from the gateways which may, but need not, be connected to the Internet. The coverage area of the radio nodes working as a single network is sometimes called a mesh cloud. Access to this mesh cloud is dependent on the radio nodes working in harmony with each other to create a radio network. A mesh network is reliable and offers redundancy. When one node can no longer operate, the rest of the nodes can still communicate with each other, directly or through one or more intermediate nodes. Wireless mesh networks can self form and self heal. Wireless mesh networks work with different wireless technologies including 802.11, 802.15, 802.16, cellular technologies and need not be restricted to any one technology or protocol.

History

Architecture

Wireless mesh architecture is a first step towards providing cost effective and low mobility over a specific coverage area. Wireless mesh infrastructure is, in effect, a network of routers minus the cabling between nodes. It's built of peer radio devices that don't have to be cabled to a wired port like traditional WLAN access points (AP) do. Mesh infrastructure carries data over large distances by splitting the distance into a series of short hops. Intermediate nodes not only boost the signal, but cooperatively pass data from point A to point B by making forwarding decisions based on their knowledge of the network, i.e. perform routing by first deriving the topology of the network.

Wireless mesh networks is a relatively "stable-topology" network except for the occasional failure of nodes or addition of new nodes. The path of traffic, being aggregated from a large number of end users, changes infrequently. Practically all the traffic in an infrastructure mesh network is either forwarded to or from a gateway, while in wireless ad hoc networks or client mesh networks the traffic flows between arbitrary pairs of nodes.

If rate of mobility among nodes are high, i.e., link breaks happen frequently, wireless mesh networks will start to break down and have low communication performance.

Management

This type of infrastructure can be decentralized (with no central server) or centrally managed (with a central server). Both are relatively inexpensive, and can be very reliable and resilient, as each node needs only transmit as far as the next node. Nodes act as routers to transmit data from nearby nodes to peers that are too far away to reach in a single hop, resulting in a network that can span larger distances. The topology of a mesh network has to be relatively stable, i.e., not too much mobility. If one node drops out of the network, due to hardware failure or any other reason, its neighbors can quickly find another route using a routing protocol.

Applications

Mesh networks may involve either fixed or mobile devices. The solutions are as diverse as communication needs, for example in difficult environments such as emergency situations, tunnels, oil rigs, battlefield surveillance, high-speed mobile-video applications on board public transport or real-time racing-car telemetry. An important possible application for wireless mesh networks is VoIP. By using a Quality of Service scheme, the wireless mesh may support local telephone calls to be routed through the mesh. Most applications in wireless mesh networks are similar to those in wireless ad hoc networks.

Some current applications:

- U.S. military forces are now using wireless mesh networking to connect their computers, mainly ruggedized laptops, in field operations.

- Electric smart meters now being deployed on residences transfer their readings from one to another and eventually to the central office for billing without the need for human meter readers or the need to connect the meters with cables.

- The laptops in the One Laptop per Child program use wireless mesh networking to enable students to exchange files and get on the Internet even though they lack wired or cell phone or other physical connections in their area.

- Google Home, Google Wi-Fi, and Google OnHub all support Wi-Fi mesh (i.e., Wi-Fi ad hoc) networking.

- The 66-satellite Iridium constellation operates as a mesh network, with wireless links between adjacent satellites. Calls between two satellite phones are routed through the mesh, from one satellite to another across the constellation, without having to go through an earth station. This makes for a smaller travel distance for the signal, reducing latency, and also allows for the constellation to operate with far fewer earth stations than would be required for 66 traditional communications satellites.

Operation

The principle is similar to the way packets travel around the wired Internet – data will hop from one device to another until it eventually reaches its destination. Dynamic routing algorithms implemented in each device allow this to happen. To implement such dynamic routing protocols, each device needs to communicate routing information to other devices in the network. Each device then determines what to do with the data it receives – either pass it on to the next device or keep it, depending on the protocol. The routing algorithm used should attempt to always ensure that the data takes the most appropriate (fastest) route to its destination.

Multi-radio Mesh

Multi-radio mesh refers to having different radios operating at different frequencies to interconnect nodes in a mesh. This means there is a unique frequency used for each wireless hop and thus a dedicated CSMA collision domain. With more radio bands, communication throughput is likely to increase as a result of more available communication channels. This is similar to providing dual or multiple radio paths to transmit and receive data.

Research Topics

One of the more often cited papers on Wireless Mesh Networks identified the following areas as open research problems in 2005

- New modulation scheme

  o In order to achieve higher transmission rate, new wideband transmission schemes other than OFDM and UWB are needed.

- Advanced antenna processing

  o Advanced antenna processing including directional, smart and multiple antenna technologies is further investigated, since their complexity and cost are still too high for wide commercialization.

- Flexible spectrum management

  o Tremendous efforts on research of frequency-agile techniques are being performed for increased efficiency.

- Cross-layer optimization

  o Cross-layer research is a popular current research topic where information is shared between different communications layers in order to increase the knowledge and current state of the network. This could enable new and more efficient protocols to be developed. A joint protocol which combines various design problems like routing, scheduling, channel assignment etc. can achieve higher performance since it is proven that these problems are strongly co-related. It is important to note that careless cross-layer design could lead to code which is difficult to maintain and extend.

- Software-defined wireless networking

  o Centralized, distributed, or hybrid? - In a new SDN architecture for WDNs is explored that eliminates the need for multi-hop flooding of route information and therefore enables WDNs to easily expand. The key idea is to split network control and data forwarding by using two separate frequency bands. The forwarding nodes and the SDN controller exchange link-state information and other network control signaling in one of the bands, while actual data forwarding takes place in the other band.

- Security

  o A WMN can be seen as a group of nodes (clients or routers) that cooperate to provide connectivity. Such an open architecture, where clients serve as routers to forward data packets, is exposed to many types of attacks that can interrupt the whole network and cause denial of service (DoS) or Distributed Denial of Service (DDoS).

Protocols

Routing Protocols

There are more than 70 competing schemes for routing packets across mesh networks. Some of these include:

- Associativity-Based Routing (ABR)

- AODV (Ad hoc On-Demand Distance Vector)

- B.A.T.M.A.N. (Better Approach To Mobile Adhoc Networking)

- Babel (protocol) (a distance-vector routing protocol for IPv6 and IPv4 with fast convergence properties)

- Dynamic NIx-Vector Routing|DNVR

- DSDV (Destination-Sequenced Distance-Vector Routing)

- DSR (Dynamic Source Routing)

- HSLS (Hazy-Sighted Link State)

- HWMP (Hybrid Wireless Mesh Protocol, the default mandatory routing protocol of IEEE 802.11s)

- *Infrastructure Wireless Mesh Protocol* (IWMP) for Infrastructure Mesh Networks by GRECO UFPB-Brazil

- OLSR (Optimized Link State Routing protocol)

- OORP (OrderOne Routing Protocol) (OrderOne Networks Routing Protocol)

- OSPF (Open Shortest Path First Routing)

- Routing Protocol for Low-Power and Lossy Networks (IETF ROLL RPL protocol, RFC 6550)

- PWRP (Predictive Wireless Routing Protocol)

- TORA (Temporally-Ordered Routing Algorithm)

- ZRP (Zone Routing Protocol)

The IEEE has developed a set of standards under the title 802.11s.

A less thorough list can be found at Ad hoc routing protocol list.

Autoconfiguration Protocols

Standard autoconfiguration protocols, such as DHCP or IPv6 stateless autoconfiguration may be used over mesh networks.

Mesh network specific autoconfiguration protocols include:

- Ad Hoc Configuration Protocol (AHCP)

- Proactive Autoconfiguration (Proactive Autoconfiguration Protocol)

- Dynamic WMN Configuration Protocol (DWCP)

Communities and Providers

- AWMN

- CUWiN

- Freifunk (DE) / FunkFeuer (AT) / OpenWireless (CH)

- Firechat

- Firetide
- Guifi.net
- Netsukuku
- Ninux (IT)

Products

- Aruba AirMesh- multiservice wireless mesh networks for outdoors
- Ruckus Mesh - Smart Mesh
- FireTide - Wireless mesh networks
- Cisco Meraki - Mesh networking - access points as gateways and repeaters
- Juniper Wireless Mesh - Wireless mesh and bridging
- OpenMesh - Open Mesh makes WiFi smarter and simpler
- StrixSystems - 802.11n powered wireless mesh
- MeshDynamics - Meshdynamics Industrial Mesh Networks
- Rajant Mesh - Kinetic Mesh Networks
- Others - list of venture backed mesh networking companies.

Municipal Wireless Network

Municipal wireless network (Municipal Wi-Fi, Muni Wi-Fi or Muni-Fi) is a citywide wireless network. This is usually done by providing municipal broadband via Wi-Fi to large parts or all of a municipal area by deploying a wireless mesh network. The typical deployment design uses hundreds of wireless access points deployed outdoors, often on poles. The operator of the network acts as a wireless internet service provider.

Overview

Municipal wireless networks go far beyond the existing piggybacking opportunities available near public libraries and some coffee shops. The basic premise of carpeting an area with wireless service in urban centers is that it is more economical to the community to provide the service as a utility rather than to have individual households and businesses pay private firms for such a service. Such networks are capable of enhancing city management and public safety, especially when used directly by city employees in the field. They can also be a social service to those who cannot afford private high-speed services. When the network service is free and a small number of clients consume a majority of the available capacity, operating and regulating the network might prove difficult.

In 2003, Verge Wireless formed an agreement with Tropos Networks to build a municipal wireless networks in the downtown area of Baton Rouge, Louisiana. Carlo MacDonald, the founder of Verge Wireless, suggested that it could provide cities a way to improve economic development and developers to build mobile applications that can make use of faster bandwidth. Verge Wireless built networks for Baton Rouge, New Orleans, and other areas. Some applications include wireless security cameras, police mug shot software, and location-based advertising.

In 2006 the US Federal Trade Commission expressed concerns about such private-public partnerships as trending towards a franchise monopoly.

In 2007, some companies with existing cell sites offered high-speed wireless services where the laptop owner purchased a PC card or adapter based on EV-DO cellular data receivers or WiMAX rather than 802.11b/g. A few high-end laptops at that time featured built-in support for these newer protocols. WiMAX is designed to implement a metropolitan area network (MAN) while 802.11 is designed to implement a wireless local area network (LAN).

Within the United States, providing a municipal wireless network is not recognized as a priority. Some have argued that the benefits of public approach may exceed the costs, similar to cable television.

Finance

The construction of such networks is a significant part of their lifetime costs. Usually, a private firm works with local government to construct a network and operate it. Financing is usually shared by both the private firm and the municipal government. Once operational, the service may be free to users via public finance or advertising, or may be a paid service. Among deployed networks, usage as measured by number of distinct users has been shown to be moderate to light. Private firms serving multiple cities sometimes maintain an account for each user, and allow the user a limited amount of mobile service in the cities covered. As of 2007 some Muni WiFi deployments are delayed as the private and public partners negotiate the business model and financing.

In such networks, radio communication is used both for the Wi-Fi service and for the "backhaul" or pathway to the Internet. This means that the nodes only need a wire for power (hence the habit of installing them on power and light utility poles). This "all radio" approach means that nodes must be within range of each other and form a contiguous pathway back to special aggregation nodes that have more traditional access to the Internet. Nodes then relay traffic, somewhat like a bucket brigade, from the laptop to the aggregation node. This limits the way in which the network can be grown incrementally: coverage starts near the aggregation point and, as the mesh grows, new coverage can only grow out from the edge of the mesh. If a new, isolated area is to be covered, then a new aggregation point must be constructed. Private firms often take a phased approach, starting with one or a few sectors of a city to demonstrate competence before making the larger investment of attempting full coverage of a city.

Google WiFi is entirely funded by Google. Despite a failed attempt to provide city-wide WiFi through a partnership with internet service provider Earthlink in 2007, the company claims that they are working to provide a wireless network for the city of San Francisco, California, although there is no specified completion date. Some other projects that are still in the planning stages have pared back their planned coverage from 100% of a municipal area to only densely commercially zoned areas. One of the most ambitious planned projects is to provide wireless service throughout Silicon Valley, but the winner of the bid seems ready to request that the 40 cities involved help cover more of the cost, which has raised concerns that the project will ultimately be too slow to market to be a success. Advances in technology in 2005–2007 may allow wireless community network projects to offer a viable alternative. Such projects have an advantage in that, as they do not have to negotiate with government entities, they have no contractual obligations for coverage. A promising example is Meraki's demonstration in San Francisco, which already claims 20,000 distinct users as of October 2007.

In 2009, Microsoft and Yahoo also provided free wireless to select regions in the United States. Yahoo's free WiFi was made available for one year to the Times Square area in New York City beginning November 10, 2009. Microsoft made free WiFi available to select airports and hotels across the United States, in exchange for one search on the Bing search engine by the user.

Potential Externalities

Unintended externalities are possible as a result of local governments providing Internet service to their constituents. A private service provider could choose to offer limited or no service to a region if that region's largest city opted to provide free Internet service, thus eliminating the potential customer base. The private sector receives no money from taxpayers, so there isn't competition. The lack of competition prevents other municipalities in that region from benefiting from the services of the private provider. The smaller public municipalities would at the same time not benefit from the free service provided by the larger city because it is designed to be subsidized by taxpayers and not concerned about the maximization of profits. The broadband provided by the government isn't largely supported to create an income on top of the private sector not being competed with enough to make a profit. Thus, making both municipal wireless networks anticompetitive.

Overuse could be another issue. If usage of the publicly provided network became heavier than existing private options network overload issues could arise, forcing the municipality to invest more heavily, thus spending more revenue, on infrastructure to maintain the existing level of service. This issue could be compounded if private providers begin exiting a market as mentioned above.

Meru Networks

Headquarters of Meru Networks

Meru Networks was a supplier of wireless local area networks (WLANs) to healthcare, enterprise, hospitality, K-12 education, higher education, and other markets. Founded in 2002 and headquartered in Sunnyvale, California, United States, the company made its initial public offering in March 2010.

History

Meru Networks was founded in 2002 to address issues with legacy wireless networking architectures that support two separate access networks: a wired network for business-specific applications and a wireless network for casual use. This causes problems ranging from co-channel interference to the inability of micro-cellular systems to scale up. Meru Networks develops and markets a virtualized wireless LAN solution that enables enterprises to migrate applications from wired networks to wireless networks and become what Meru refers to as the "All-wireless enterprise." The company uses an approach to wireless networking that employs virtualization technology to create a self-monitoring wireless network that provides access to applications, improved application performance, and a greater ability to run converged applications, such as voice, video and data, over a wireless network.

The company's current products address the IEEE 802.11ac and 802.11n wireless networking standards, The company focuses on a "Virtual Cell" approach to Wi-Fi. Under a single service assurance platform, it aggregates access points and the controller needed to manage them. This simplifies the management of access points, cuts the number of access points needed on a wireless network, and eliminates bandwidth contention issues.

Timeline

- February 2002: Meru Networks founded by Dr. Vaduvur Bharghavan, Srinath Sarang, Sung-Wook Han, Joe Epstein,

- June 2005: Company receives $12 M in Series C funding

- May 2006: Company receives $25 M in Series D funding

- January 2007: Company receives $27.6 M in unattributed funding

- January 2009: Company adds Ihab Abu-Hakima as President and CEO

- April 2009: Company receives $30 M in Series E funding

- August 2009: Company receives an additional $57 M in Series E funding

- March 2010: Company goes public with stock symbol NASDAQ:MERU

- May 2011: Dr. Vaduvur Bharghavan founds another company, becomes Advisory CTO

- March 2012: Bami Bastani appointed as President and CEO

Main Products and Services

Software

- Meru System Director Operating System. Runs on all Meru controllers and access points, implementing its virtualized wireless LAN technology. As an operating system, System Director runs other applications to deal with the specific requirements of the enterprise. Each application module is integrated with System Director, using its low-level control over the radio frequencies.

- Meru Identity Manager. Allows businesses to provide access to thousands of Wi-Fi devices in the "bring your own device" (BYOD) workplace.

- Meru Service Assurance Manager. Provides network-wide operations monitoring and diagnostics for wireless networks and the applications that run over them.

- Meru E(z)RF Network Manager. Manages multiple controllers and thousands of access points providing real-time location tracking and location firewall.

- Meru Spectrum Manager. A spectrum analysis solution.

Hardware

- Access Points. Built from the radio layer up to support virtualized wireless LANs and, with Meru software, to allow enterprise IT groups to replace wired Ethernet switches at the network edge.

- Controllers. Supporting from five to thousands of access points, Meru WLAN controllers synchronize these points to avoid interference and govern all traffic on the network to support more users and backhaul capacity.

- Appliances. The Meru Services Appliance family is an extensible platform that provides a set of applications for management of 802.11x networks. The appliance collects a range of data from Meru access points, storing it in a common database, where it is used by the higher-level management applications.

Products and services are sold through value-added resellers (VARs) and distributors.

Recognition

2011 honors include:

- 2011 ITP Technology Best Wireless Solution

- 2011 Health Management Technology Coolest Products

Markets

The company focuses on the education, healthcare and hospitality markets. Approximately three-quarters of its global sales come from the Americas, and among its 12,500 customers worldwide are the CME Group and Hellmann Worldwide Logistics.

While the company's main emphasis is on the markets noted above, they are not its exclusive focus. In December 2011, for example, Meru announced that it was working with Ayacht Technology Solutions to install a new Wi-Fi network at historic, 37,000-seat Fenway Park, the home of Major League Baseball's Boston Red Sox.

Acquisitions

On September 8, 2011, Meru announced that it had acquired access control specialist Identity Networks. Founded in 2006 and based in Manchester, UK, Identity Networks develops and markets tools that offer provisioning and notification services that give guests access credentials by way of email, or SMS, and also provide reporting, auditing, and policy customization.

Acquisition

On May 27, 2015, Fortinet agreed to acquire Meru Networks for $44 million in cash. Fortinet will integrate the Meru Networks team and operations into their organization.

Node (Networking)

In communication networks, a node (Latin *nodus*, 'knot') is either a redistribution point (e.g. data communications equipment), or a communication endpoint (e.g. data terminal equipment). The definition of a node depends on the network and protocol layer referred to. A physical network node is an active electronic device that is attached to a network, and is capable of creating, receiving, or transmitting information over a communications channel. A passive distribution point such as a distribution frame or patch panel is consequently not a node.

Computer Network Nodes

In data communication, a physical network node may either be a data communication equipment (DCE) such as a modem, hub, bridge or switch; or a data terminal equipment (DTE) such as a digital telephone handset, a printer or a host computer, for example a router, a workstation or a server.

If the network in question is a LAN or WAN, every LAN or WAN node (that are at least data link layer devices) must have a MAC address, typically one for each network interface controller it possesses. Examples are computers, packet switches, xDSL modems (with Ethernet interface) and wireless LAN access points. Note that a hub constitutes a physical network node, but does not constitute a LAN network node, since a hubbed network logically is a bus network. Analogously, a repeater or PSTN modem (with serial interface) is a physical network node but not a LAN node in this sense.

If the network in question is the Internet or an Intranet, many physical network nodes are host computers, also known as Internet nodes, identified by an IP address, and all hosts are physical network nodes. However, some datalink layer devices such as switches, bridges and WLAN access points do not have an IP host address (except sometimes for administrative purposes), and are not considered to be Internet nodes or hosts, but as physical network nodes and LAN nodes.

Telecommunication Network Nodes

In the fixed telephone network, a node may be a public or private telephone exchange, a remote concentrator or a computer providing some intelligent network service. In cellular communication, switching points and databases such as the Base station controller, Home Location Register, Gateway GPRS Support Node (GGSN) and Serving GPRS Support Node (SGSN) are examples of nodes. Cellular network base stations are not considered to be nodes in this context.

In cable television systems (CATV), this term has assumed a broader context and is generally associated with a fiber optic node. This can be defined as those homes or businesses within a specific geographic area that are served from a common fiber optic receiver. A fiber optic node is generally described in terms of the number of "homes passed" that are served by that specific fiber node.

Distributed System Nodes

If the network in question is a distributed system, the nodes are clients, servers or peers. A peer may sometimes serve as client, sometimes server. In a peer-to-peer or overlay network, nodes that actively route data for the other networked devices as well as themselves are called supernodes.

Distributed systems may sometimes use *virtual nodes* so that the system is not oblivious to the heterogeneity of the nodes. This issue is addressed with special algorithms, like consistent hashing, as it is the case in Amazon's.

End Node in Cloud Computing

Within a vast computer network, the individual computers on the periphery of the network, those that do not also connect other networks, and those that often connect transiently to one or more clouds are called end nodes. Typically, within the cloud computing construct, the individual user / customer computer that connects into one well-managed cloud is called an end node. Since these computers are a part of the network yet unmanaged by the cloud's host, they present significant risks to the entire cloud. This is called the End Node Problem. There are several means to remedy this problem but all require instilling trust in the end node computer.

Free-space Optical Communication

Free-space optical communication (FSO) is an optical communication technology that uses light propagating in free space to wirelessly transmit data for telecommunications or computer networking. "Free space" means air, outer space, vacuum, or something similar. This contrasts with using solids such as optical fiber cable or an optical transmission line. The technology is useful where the physical connections are impractical due to high costs or other considerations.

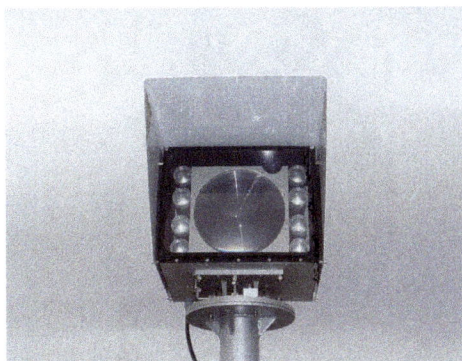

An 8-beam free space optics laser link, rated for 1 Gbit/s. The receptor is the large disc in the middle, the transmitters the smaller ones. At the top right corner is a monocular for assisting the alignment of the two heads.

History

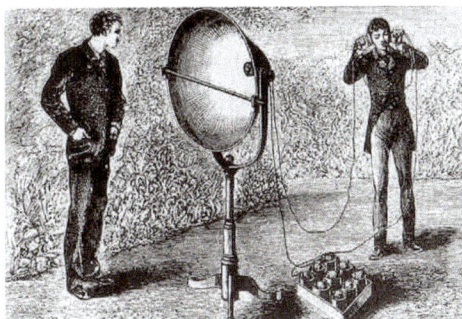

A photophone receiver and headset, one half of Bell and Tainter's optical telecommunication system of 1880.

Optical communications, in various forms, have been used for thousands of years. The Ancient Greeks used a coded alphabetic system of signalling with torches developed by Cleoxenus, Democleitus and Polybius. In the modern era, semaphores and wireless solar telegraphs called heliographs were developed, using coded signals to communicate with their recipients.

In 1880, Alexander Graham Bell and his assistant Charles Sumner Tainter created the photophone, at Bell's newly established Volta Laboratory in Washington, DC. Bell considered it his most important invention. The device allowed for the transmission of sound on a beam of light. On June 3, 1880, Bell conducted the world's first wireless telephone transmission between two buildings, some 213 meters (700 feet) apart.

Its first practical use came in military communication systems many decades later, first for optical telegraphy. German colonial troops used heliograph telegraphy transmitters during the Herero and Namaqua genocide starting in 1904, in German South-West Africa (today's Namibia) as did British, French, US or Ottoman signals.

WW I German Blinkgerät

During the trench warfare of World War I when wire communications were often cut, German signals used three types of optical Morse transmitters called *Blinkgerät*, the intermediate type for distances of up to 4 km (2.5 miles) at daylight and of up to 8 km (5 miles) at night, using red filters for undetected communications. Optical telephone communications were tested at the end of the war, but not introduced at troop level. In addition, special blinkgeräts were used for communication with airplanes, balloons, and tanks, with varying success.

A major technological step was to replace the Morse code by modulating optical waves in speech transmission. Carl Zeiss, Jena developed the *Lichtsprechgerät 80/80* (literal translation: optical speaking device) that the German army used in their World War II anti-aircraft defense units, or in bunkers at the Atlantic Wall.

The invention of lasers in the 1960s, revolutionized free space optics. Military organizations were particularly interested and boosted their development. However the technology lost market momentum when the installation of optical fiber networks for civilian uses was at its peak.

Many simple and inexpensive consumer remote controls use low-speed communication using infrared (IR) light. This is known as consumer IR technologies.

Usage and Technologies

Free-space point-to-point optical links can be implemented using infrared laser light, although low-data-rate communication over short distances is possible using LEDs. Infrared Data Association (IrDA) technology is a very simple form of free-space optical communications. On the communications side the FSO technology is considered as a part of the Optical Wireless Communications applications. Free-space optics can be used for communications between spacecraft.

Current Market Demands

The demand for a high-speed (10+ Gbit/s) and long range (3–5 km) FSO system is apparent in the market place.

- In 2008, MRV Communications introduced a free-space optics (FSO)-based system with a data rate of 10 Gbit/s initially claiming a distance of 2 km at high availability. This equipment is no longer available; before end-of-life, the product's useful distance was changed down to 350 m.

- In 2013, the company MOSTCOM started to serially produce a new wireless communication system that also had a data rate of 10 Gbit/s as well as an improved range of up to 2.5 km, but to get to 99.99% uptime the designers used an RF hybrid solution, meaning the data rate drops to extremely low levels during atmospheric disturbances (typically down to 10 Mbit/s). In April 2014, the company with Scientific and Technological Centre "Fiord" demonstrated the transmission speed 30 Gbit/s under "laboratory conditions".

- LightPointe offers many similar hybrid solutions to MOSTCOM's offering.

Useful Distances

The reliability of FSO units has always been a problem for commercial telecommunications. Consistently, studies find too many dropped packets and signal errors over small ranges (400 to 500 meters). This is from both independent studies, such as in the Czech republic, as well as formal internal nationwide studies, such as one conducted by MRV FSO staff. Military based studies consistently produce longer estimates for reliability, projecting the maximum range for terrestrial links is of the order of 2 to 3 km (1.2 to 1.9 mi). All studies agree the stability and quality of the link is highly dependent on atmospheric factors such as rain, fog, dust and heat.

Extending the Useful Distance

The main reason terrestrial communications have been limited to non-commercial telecommunications functions is fog. Fog consistently keeps FSO laser links over 500 meters from achieving a year-round bit error rate of 1 per 100,000. Several entities are continually attempting to overcome these key disadvantages to FSO communications and field a system with a better quality of service. DARPA has sponsored over US$130 million in research towards this effort, with the ORCA and ORCLE programs.

DARPA ORCA Official Concept Art created c. 2008

Other non-government groups are fielding tests to evaluate different technologies that some claim have the ability to address key FSO adoption challenges. As of October 2014, none have fielded a working system that addresses the most common atmospheric events.

FSO research from 1998–2006 in the private sector totaled $407.1 million, divided primarily among four start-up companies. All four failed to deliver products that would meet telecommunications quality and distance standards:

- Terabeam received approximately $226 million in funding. AT&T and Lucent backed this attempt. The work ultimately failed, and the company reorganized in 2004.

- AirFiber received $96.1 million in funding, and never solved the weather issue. They sold out to MRV communications in 2003, and MRV sold their FSO units until 2012 when the end-of-life was abruptly announced for the Terescope series.

- LightPointe Communications received $76 million in start-up funds, and eventually reorganized to sell hybrid FSO-RF units to overcome the weather-based challenges.

- The Maxima Corporation published its operating theory in Science (magazine), and received $9 million in funding before permanently shutting down. No known spin-off or purchase followed this effort.

- Wireless Excellence developed and launched CableFree UNITY solutions that combine FSO with Millimeter Wave and Radio technologies to extend distance, capacity and availability, with a goal of making FSO a more useful and practical technology.

One private company published a paper on November 20, 2014, claiming they had achieved commercial reliability (99.999% availability) in extreme fog. There is no indication this product is currently commercially available.

Extraterrestrial

The massive advantages of laser communication in space have multiple space agencies racing to develop a stable space communication platform, with many significant demonstrations and achievements. As of 18 December 2014, *no laser communication system is in use in space.*

Demonstrations in Space:

The first gigabit laser-based communication was achieved by the European Space Agency and called the European Data Relay System (EDRS) on November 28, 2014. The initial images have just been demonstrated, and a working system is expected to be in place in the 2015–2016 time frame.

NASA's OPALS announced a breakthrough in space-to-ground communication December 9, 2014, uploading 175 megabytes in 3.5 seconds. Their system is also able to re-acquire tracking after the signal was lost due to cloud cover.

In January 2013, NASA used lasers to beam an image of the Mona Lisa to the Lunar Reconnaissance Orbiter roughly 390,000 km (240,000 mi) away. To compensate for atmospheric interference, an error correction code algorithm similar to that used in CDs was implemented.

A two-way distance record for communication was set by the Mercury laser altimeter instrument aboard the MESSENGER spacecraft, and was able to communicate across a distance of 24 million km (15 million miles), as the craft neared Earth on a fly-by in May, 2005. The previous record had been set with a one-way detection of laser light from Earth, by the Galileo probe, of 6 million km in 1992. Quote from Laser Communication in Space Demonstrations (EDRS).

LEDs

RONJA is a free implementation of FSO using high-intensity LEDs

In 2001, Twibright Labs released Ronja Metropolis, an open source DIY 10 Mbit/s full duplex LED FSO over 1.4 km In 2004, a Visible Light Communication Consortium was formed in Japan. This was based on work from researchers that used a white LED-based space lighting system for indoor local area network (LAN) communications. These systems present advantages over traditional UHF RF-based systems from improved isolation between systems, the size and cost of receivers/transmitters, RF licensing laws and by combining space lighting and communication into the same system. In January 2009, a task force for visible light communication was formed by the Institute of Electrical and Electronics Engineers working group for wireless personal area network standards known as IEEE 802.15.7. A trial was announced in 2010, in St. Cloud, Minnesota.

Amateur radio operators have achieved significantly farther distances using incoherent sources of light from high-intensity LEDs. One reported 173 miles (278 km) in 2007. However, physical limitations of the equipment used limited bandwidths to about 4 kHz. The high sensitivities required of the detector to cover such distances made the internal capacitance of the photodiode used a dominant factor in the high-impedance amplifier which followed it, thus naturally forming a low-pass filter with a cut-off frequency in the 4 kHz range. From the other side use of lasers radiation source allows to reach very high data rates which are comparable to fiber communications.

Projected data rates and future data rate claims vary. A low-cost white LED (GaN-phosphor) which could be used for space lighting can typically be modulated up to 20 MHz. Data rates of over 100 Mbit/s can be easily achieved using efficient modulation schemes and Siemens claimed to have achieved over 500 Mbit/s in 2010. Research published in 2009, used a similar system for traffic control of automated vehicles with LED traffic lights.

In September 2013, pureLiFi, the Edinburgh start-up working on Li-Fi, also demonstrated high speed point-to-point connectivity using any off-the-shelf LED light bulb. In previous work, high bandwidth specialist LEDs have been used to achieve the high data rates. The new system, the Li-1st, maximizes the available optical bandwidth for any LED device, thereby reducing the cost and improving the performance of deploying indoor FSO systems.

Engineering Details

Typically, best use scenarios for this technology are:

- LAN-to-LAN connections on campuses at Fast Ethernet or Gigabit Ethernet speeds

- LAN-to-LAN connections in a city, a metropolitan area network

- To cross a public road or other barriers which the sender and receiver do not own

- Speedy service delivery of high-bandwidth access to optical fiber networks

- Converged Voice-Data-Connection

- Temporary network installation (for events or other purposes)

- Reestablish high-speed connection quickly (disaster recovery)

- As an alternative or upgrade add-on to existing wireless technologies

  o Especially powerful in combination with auto aiming systems, this way you could power moving cars or you can power your laptop while you move or use auto-aiming nodes to create a network with other nodes.

- As a safety add-on for important fiber connections (redundancy)

- For communications between spacecraft, including elements of a satellite constellation

- For inter- and intra-chip communication

The light beam can be very narrow, which makes FSO hard to intercept, improving security. In any case, it is comparatively easy to encrypt any data traveling across the FSO connection for additional security. FSO provides vastly improved electromagnetic interference (EMI) behavior compared to using microwaves.

Technical Advantages

- Ease of deployment

- Can be used to power devices

- License-free long-range operation (in contrast with radio communication)

- High bit rates

- Low bit error rates

- Immunity to electromagnetic interference

- Full duplex operation

- Protocol transparency

- Increased security when working with narrow beam(s)

- No Fresnel zone necessary

- Reference open source implementation

Range Limiting Factors

For terrestrial applications, the principal limiting factors are:

- Fog (10 to ~100 dB/km attenuation)

- Beam dispersion

- Atmospheric absorption

- Rain

- Snow

- Terrestrial scintillation

- Interference from background light sources (including the Sun)

- Shadowing

- Pointing stability in wind

- Pollution / smog

These factors cause an attenuated receiver signal and lead to higher bit error ratio (BER). To overcome these issues, vendors found some solutions, like multi-beam or multi-path architectures, which use more than one sender and more than one receiver. Some state-of-the-art devices also have larger fade margin (extra power, reserved for rain, smog, fog). To keep an eye-safe environment, good FSO systems have a limited laser power density and support laser classes 1 or 1M. Atmospheric and fog attenuation, which are exponential in nature, limit practical range of FSO devices to several kilometres.However the Free space optics, based on 1550 nm wavelength, have considerably lower optical loss than Free space optics, using 830 nm wavelength, in dense fog conditions. FSO using wavelength 1550 nm system are capable of transmitting several times higher power than systems with 850 nm and are at the same time safe to the human eye (1M class). Some of Free space optics, as for example EC SYSTEM, use in additional to that Automatic Gain Control in-built to regulate the transmit power of the laser diode depending on the link quality, which is constantly monitored and allowe increase power of laser transmitter to ensure higher reliability of connection in the bad weather conditions.

Wireless Network Interface Controller

A wireless network interface controller (WNIC) is a network interface controller which connects to a wireless radio-based computer network, rather than a wired network, such as Token Ring or Ethernet. A WNIC, just like other NICs, works on the Layer 1 and

Layer 2 of the OSI Model. This card uses an antenna to communicate via microwave radiation. A WNIC in a desktop computer is traditionally connected using the PCI bus. Other connectivity options are USB and PC card. Integrated WNICs are also available, (typically in Mini PCI/PCI Express Mini Card form).

A wireless network interface device with a USB interface and internal antenna

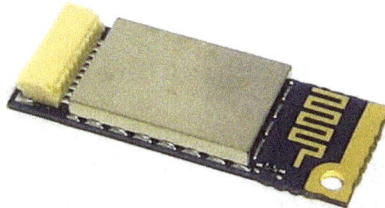

A Bluetooth interface card

Early wireless network interface controllers were commonly implemented on expansion cards that plugged into a computer bus. The low cost and ubiquity of the Wi-Fi standard means that many newer mobile computers have a wireless network interface built into the motherboard.

The term is usually applied to IEEE 802.11 adapters; it may also apply to a NIC using protocols other than 802.11, such as one implementing Bluetooth connections.

Modes of Operation

An 802.11 WNIC can operate in two modes known as *infrastructure mode* and *ad hoc mode*:

Infrastructure Mode

In an infrastructure mode network the WNIC needs a wireless access point: all data is transferred using the access point as the central hub. All wireless nodes in an infrastructure mode network connect to an access point. All nodes connecting to the access point must have the same service set identifier (SSID) as the access point, and if a kind of wireless security is enabled on the access point (such as WEP or WPA), they must share the same keys or other authentication parameters.

Ad Hoc Mode

In an ad hoc mode network the WNIC does not require an access point, but rather can interface with all other wireless nodes directly. All the nodes in an ad hoc network must have the same channel and SSID.

Specifications

The IEEE 802.11 standard sets out low-level specifications for how all 802.11 wireless networks operate. Earlier 802.11 interface controllers are usually only compatible with earlier variants of the standard, while newer cards support both current and old standards.

Specifications commonly used in marketing materials for WNICs include:

- Wireless data transfer rates (measured in Mbit/s); these range from 2 Mbit/s to 54 Mbit/s.

- Wireless transmit power (measured in dBm)

- Wireless network standards (may include standards such as 802.11b, 802.11g, 802.11n, etc.) 802.11g offers data transfer speeds equivalent to 802.11a – up to 54 Mbit/s – and the wider 300-foot (91 m) range of 802.11b, and is backward compatible with 802.11b.

Most Bluetooth cards do not implement any form of the 802.11 standard.

Range

Wireless range may be substantially affected by objects in the way of the signal and by the quality of the antenna. Large electrical appliances, such as refrigerators, fuse boxes, metal plumbing, and air conditioning units can impede a wireless network signal. The theoretical maximum range of IEEE 802.11 is only reached under ideal circumstances and true effective range is typically about half of the theoretical range. Specifically, the maximum throughput speed is only achieved at extremely close range (less than 25 feet (7.6 m) or so); at the outer reaches of a device's effective range, speed may decrease to around 1 Mbit/s before it drops out altogether. The reason is that wireless devices dynamically negotiate the top speed at which they can communicate without dropping too many data packets.

FullMAC and SoftMAc Devices

In an 802.11 WNIC, the *MAC Sublayer Management Entity* (MLME) can be implemented either in the NIC's hardware or firmware, or in host-based software that is executed on the main CPU. A WNIC that implements the MLME function in hardware or firmware is called a *FullMAC* WNIC or a *HardMAC* NIC and a NIC that implements it in host software is called a *SoftMAC* NIC.

A FullMAC device hides the complexity of the 802.11 protocol from the main CPU, instead providing an 802.3 (Ethernet) interface; a SoftMAC design implements only the timing-critical part of the protocol in hardware/firmware and the rest on the host.

FullMAC chips are typically used in mobile devices because:

- they are easier to integrate in complete products

- power is saved by having a specialized CPU perform the 802.11 processing;

- the chip vendor has tighter control of the MLME.

Popular example of FullMAC chips is the one implemented on the Raspberry Pi 3.

Linux kernel's *mac80211* framework provides capabilities for SoftMAC devices and additional capabilities (such as mesh networking, which is known as the IEEE 802.11s standard) for devices with limited functionality.

FreeBSD also supports SoftMAC drivers.

Fixed Wireless

CableFree Fixed Wireless Microwave Backhaul links deployed for mobile operators in the Middle East. These microwave links typically carry a mix of Ethernet /IP, TDM (Nx E1) and SDH traffic to connect sites with high capacity. Such microwave links used to carry 2xE1 (4Mbit/s) now carry 800Mbit/s or more, using modern 1024QAM or higher modulation schemes.

Fixed wireless is the operation of wireless devices or systems used to connect two fixed locations (e.g., building to building or tower to building) with a radio or other wireless link, such as laser bridge. Usually, fixed wireless is part of a wireless LAN infrastructure. The purpose of a fixed wireless link is to enable data communications between the two sites or buildings. Fixed wireless data (FWD) links are often a cost-effective alternative to leasing fiber or installing cables between the buildings.

The point-to-point signal transmissions occur through the air over a terrestrial microwave platform rather than through copper or optical fiber; therefore, fixed wireless does not require satellite feeds or local telephone service. The advantages of fixed wireless include the ability to connect with users in remote areas without the need for laying new cables and the capacity for broad bandwidth that is not impeded by fiber or cable capacities. Fixed wireless devices usually derive their electrical power from the public utility mains, unlike mobile wireless or portable wireless devices which tend to be battery powered.

Antennas

Fixed wireless services typically use a directional radio antenna on each end of the signal (e.g., on each building). These antennas are generally larger than those seen in Wi-Fi setups and are designed for outdoor use. Several types of radio antennas are available that accommodate various weather conditions, signal distances and bandwidths. They are usually selected to make the beam as narrow as possible and thus focus transmit power to their destination, increasing reliability and reducing the chance of eavesdropping or data injection. The links are usually arranged as a point-to-point setup to permit the use of these antennas. This also permits the link to have better speed and or better reach for the same amount of power.

These antennas are typically designed to be used in the unlicensed ISM band radio frequency bands (900 MHz, 1.8 GHz, 2.4 GHz and 5 GHz), however, in most commercial installations, licensed frequencies may be used to ensure quality of service (QoS) or to provide higher connection speeds.

Fixed Wireless Broadband

With the growing infrastructure of the GSM wireless networks, fixed wireless has also become a viable solution for broadband access. Using the 3G speed and reliability, businesses and homes can use fixed-wireless antenna technology to access broadband Internet and Layer 2 networks using fixed wireless broadband. Because of the redundancy and saturation of the GSM network, antennas that can aggregate signal from multiple carriers are able to offer fail-over and redundancy for connectivity not otherwise afforded by wired connections. In rural areas where wired infrastructure is not yet available, fixed-wireless broadband has become a viable option for Internet access.

Wireless Security

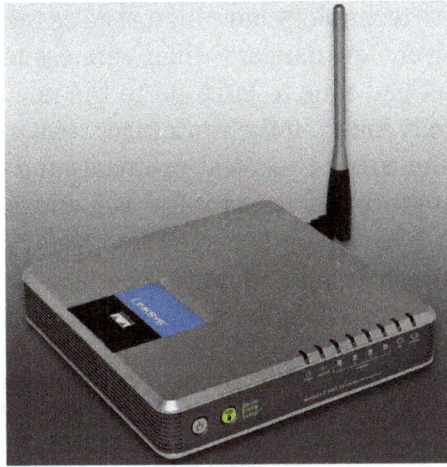

An example wireless router, that can implement wireless security features

Wireless security is the prevention of unauthorized access or damage to computers using wireless networks. The most common types of wireless security are Wired Equivalent Privacy (WEP) and Wi-Fi Protected Access (WPA). WEP is a notoriously weak security standard. The password it uses can often be cracked in a few minutes with a basic laptop computer and widely available software tools. WEP is an old IEEE 802.11 standard from 1999, which was outdated in 2003 by WPA, or Wi-Fi Protected Access. WPA was a quick alternative to improve security over WEP. The current standard is WPA2; some hardware cannot support WPA2 without firmware upgrade or replacement. WPA2 uses an encryption device that encrypts the network with a 256-bit key; the longer key length improves security over WEP.

Security settings panel for a DD-WRT router

Many laptop computers have wireless cards pre-installed. The ability to enter a network while mobile has great benefits. However, wireless networking is prone to some security issues. Hackers have found wireless networks relatively easy to break into, and even use wireless technology to hack into wired networks. As a result, it is very important that enterprises define effective wireless security policies that guard against unauthorized access to important resources. Wireless Intrusion Prevention Systems

(WIPS) or Wireless Intrusion Detection Systems (WIDS) are commonly used to enforce wireless security policies.

The risks to users of wireless technology have increased as the service has become more popular. There were relatively few dangers when wireless technology was first introduced. Hackers had not yet had time to latch on to the new technology, and wireless networks were not commonly found in the work place. However, there are many security risks associated with the current wireless protocols and encryption methods, and in the carelessness and ignorance that exists at the user and corporate IT level. Hacking methods have become much more sophisticated and innovative with wireless access. Hacking has also become much easier and more accessible with easy-to-use Windows- or Linux-based tools being made available on the web at no charge.

Some organizations that have no wireless access points installed do not feel that they need to address wireless security concerns. In-Stat MDR and META Group have estimated that 95% of all corporate laptop computers that were planned to be purchased in 2005 were equipped with wireless cards. Issues can arise in a supposedly non-wireless organization when a wireless laptop is plugged into the corporate network. A hacker could sit out in the parking lot and gather information from it through laptops and/or other devices, or even break in through this wireless card–equipped laptop and gain access to the wired network.

Background

Anyone within the geographical network range of an open, unencrypted wireless network can "sniff", or capture and record, the traffic, gain unauthorized access to internal network resources as well as to the internet, and then use the information and resources to perform disruptive or illegal acts. Such security breaches have become important concerns for both enterprise and home networks.

If router security is not activated or if the owner deactivates it for convenience, it creates a free hotspot. Since most 21st-century laptop PCs have wireless networking built in, they don't need a third-party adapter such as a PCMCIA Card or USB dongle. Built-in wireless networking might be enabled by default, without the owner realizing it, thus broadcasting the laptop's accessibility to any computer nearby.

Modern operating systems such as Linux, macOS, or Microsoft Windows make it fairly easy to set up a PC as a wireless LAN "base station" using Internet Connection Sharing, thus allowing all the PCs in the home to access the Internet through the "base" PC. However, lack of knowledge among users about the security issues inherent in setting up such systems often may allow others nearby access to the connection. Such "piggybacking" is usually achieved without the wireless network operator's knowledge; it may even be without the knowledge of the intruding user if their computer automatically selects a nearby unsecured wireless network to use as an access point.

The Threat Situation

Wireless security is just an aspect of computer security; however, organizations may be particularly vulnerable to security breaches caused by rogue access points.

If an employee (trusted entity) brings in a wireless router and plugs it into an unsecured switchport, the entire network can be exposed to anyone within range of the signals. Similarly, if an employee adds a wireless interface to a networked computer using an open USB port, they may create a breach in network security that would allow access to confidential materials. However, there are effective countermeasures (like disabling open switchports during switch configuration and VLAN configuration to limit network access) that are available to protect both the network and the information it contains, but such countermeasures must be applied uniformly to all network devices.

Threats and Vulnerabilites in an Industrial (M2M) Context

Due to its availability and low cost, the use of wireless communication technologies increases in domains beyond the originally intended usage areas, e.g. M2M communication in industrial applications. Such industrial applications often have specific security requirements. Hence, it is important to understand the characteristics of such applications and evaluate the vulnerabilities bearing the highest risk in this context. Evaluation of these vulnerabilities and the resulting vulnerability catalogs in an industrial context when considering WLAN, NFC and ZigBee are available.

The Mobility Advantage

Wireless networks are very common, both for organizations and individuals. Many laptop computers have wireless cards pre-installed. The ability to enter a network while mobile has great benefits. However, wireless networking is prone to some security issues. Hackers have found wireless networks relatively easy to break into, and even use wireless technology to hack into wired networks. As a result, it is very important that enterprises define effective wireless security policies that guard against unauthorized access to important resources. Wireless Intrusion Prevention Systems (WIPS) or Wireless Intrusion Detection Systems (WIDS) are commonly used to enforce wireless security policies.

The Air Interface and Link Corruption Risk

There were relatively few dangers when wireless technology was first introduced, as the effort to maintain the communication was high and the effort to intrude is always higher. The variety of risks to users of wireless technology have increased as the service has become more popular and the technology more commonly available. Today there are a great number of security risks associated with the current wireless protocols and encryption methods, as carelessness and ignorance exists at the user and corporate IT level. Hacking methods have become much more sophisticated and innovative with wireless.

Modes of Unauthorized Access

The modes of unauthorised access to links, to functions and to data is as variable as the respective entities make use of program code. There does not exist a full scope model of such threat. To some extent the prevention relies on known modes and methods of attack and relevant methods for suppression of the applied methods. However, each new mode of operation will create new options of threatening. Hence prevention requires a steady drive for improvement. The described modes of attack are just a snapshot of typical methods and scenarios where to apply.

Accidental Association

Violation of the security perimeter of a corporate network can come from a number of different methods and intents. One of these methods is referred to as "accidental association". When a user turns on a computer and it latches on to a wireless access point from a neighboring company's overlapping network, the user may not even know that this has occurred. However, it is a security breach in that proprietary company information is exposed and now there could exist a link from one company to the other. This is especially true if the laptop is also hooked to a wired network.

Accidental association is a case of wireless vulnerability called as "mis-association". Mis-association can be accidental, deliberate (for example, done to bypass corporate firewall) or it can result from deliberate attempts on wireless clients to lure them into connecting to attacker's APs.

Malicious Association

"Malicious associations" are when wireless devices can be actively made by attackers to connect to a company network through their laptop instead of a company access point (AP). These types of laptops are known as "soft APs" and are created when a cyber criminal runs some software that makes his/her wireless network card look like a legitimate access point. Once the thief has gained access, he/she can steal passwords, launch attacks on the wired network, or plant trojans. Since wireless networks operate at the Layer 2 level, Layer 3 protections such as network authentication and virtual private networks (VPNs) offer no barrier. Wireless 802.1x authentications do help with some protection but are still vulnerable to hacking. The idea behind this type of attack may not be to break into a VPN or other security measures. Most likely the criminal is just trying to take over the client at the Layer 2 level.

Ad hoc Networks

Ad hoc networks can pose a security threat. Ad hoc networks are defined as [peer to peer] networks between wireless computers that do not have an access point in between them. While these types of networks usually have little protection, encryption methods can be used to provide security.

The security hole provided by Ad hoc networking is not the Ad hoc network itself but the bridge it provides into other networks, usually in the corporate environment, and the unfortunate default settings in most versions of Microsoft Windows to have this feature turned on unless explicitly disabled. Thus the user may not even know they have an unsecured Ad hoc network in operation on their computer. If they are also using a wired or wireless infrastructure network at the same time, they are providing a bridge to the secured organizational network through the unsecured Ad hoc connection. Bridging is in two forms. A direct bridge, which requires the user actually configure a bridge between the two connections and is thus unlikely to be initiated unless explicitly desired, and an indirect bridge which is the shared resources on the user computer. The indirect bridge may expose private data that is shared from the user's computer to LAN connections, such as shared folders or private Network Attached Storage, making no distinction between authenticated or private connections and unauthenticated Ad-Hoc networks. This presents no threats not already familiar to open/public or unsecured wifi access points, but firewall rules may be circumvented in the case of poorly configured operating systems or local settings.

Non-traditional Networks

Non-traditional networks such as personal network Bluetooth devices are not safe from hacking and should be regarded as a security risk. Even barcode readers, handheld PDAs, and wireless printers and copiers should be secured. These non-traditional networks can be easily overlooked by IT personnel who have narrowly focused on laptops and access points.

Identity Theft (MAC Spoofing)

Identity theft (or MAC spoofing) occurs when a hacker is able to listen in on network traffic and identify the MAC address of a computer with network privileges. Most wireless systems allow some kind of MAC filtering to allow only authorized computers with specific MAC IDs to gain access and utilize the network. However, programs exist that have network "sniffing" capabilities. Combine these programs with other software that allow a computer to pretend it has any MAC address that the hacker desires, and the hacker can easily get around that hurdle.

MAC filtering is effective only for small residential (SOHO) networks, since it provides protection only when the wireless device is "off the air". Any 802.11 device "on the air" freely transmits its unencrypted MAC address in its 802.11 headers, and it requires no special equipment or software to detect it. Anyone with an 802.11 receiver (laptop and wireless adapter) and a freeware wireless packet analyzer can obtain the MAC address of any transmitting 802.11 within range. In an organizational environment, where most wireless devices are "on the air" throughout the active working shift, MAC filtering provides only a false sense of security since it prevents only "casual" or unintended connections to the organizational infrastructure and does nothing to prevent a directed attack.

Man-in-the-middle Attacks

A man-in-the-middle attacker entices computers to log into a computer which is set up as a soft AP (Access Point). Once this is done, the hacker connects to a real access point through another wireless card offering a steady flow of traffic through the transparent hacking computer to the real network. The hacker can then sniff the traffic. One type of man-in-the-middle attack relies on security faults in challenge and handshake protocols to execute a "de-authentication attack". This attack forces AP-connected computers to drop their connections and reconnect with the hacker's soft AP (disconnects the user from the modem so they have to connect again using their password which one can extract from the recording of the event). Man-in-the-middle attacks are enhanced by software such as LANjack and AirJack which automate multiple steps of the process, meaning what once required some skill can now be done by script kiddies. Hotspots are particularly vulnerable to any attack since there is little to no security on these networks.

Denial of Service

A Denial-of-Service attack (DoS) occurs when an attacker continually bombards a targeted AP (Access Point) or network with bogus requests, premature successful connection messages, failure messages, and/or other commands. These cause legitimate users to not be able to get on the network and may even cause the network to crash. These attacks rely on the abuse of protocols such as the Extensible Authentication Protocol (EAP).

The DoS attack in itself does little to expose organizational data to a malicious attacker, since the interruption of the network prevents the flow of data and actually indirectly protects data by preventing it from being transmitted. The usual reason for performing a DoS attack is to observe the recovery of the wireless network, during which all of the initial handshake codes are re-transmitted by all devices, providing an opportunity for the malicious attacker to record these codes and use various cracking tools to analyze security weaknesses and exploit them to gain unauthorized access to the system. This works best on weakly encrypted systems such as WEP, where there are a number of tools available which can launch a dictionary style attack of "possibly accepted" security keys based on the "model" security key captured during the network recovery.

Network Injection

In a network injection attack, a hacker can make use of access points that are exposed to non-filtered network traffic, specifically broadcasting network traffic such as "Spanning Tree" (802.1D), OSPF, RIP, and HSRP. The hacker injects bogus networking re-configuration commands that affect routers, switches, and intelligent hubs. A whole network can be brought down in this manner and require rebooting or even reprogramming of all intelligent networking devices.

Caffe Latte Attack

The Caffe Latte attack is another way to defeat WEP. It is not necessary for the attacker to be in the area of the network using this exploit. By using a process that targets the Windows wireless stack, it is possible to obtain the WEP key from a remote client. By sending a flood of encrypted ARP requests, the assailant takes advantage of the shared key authentication and the message modification flaws in 802.11 WEP. The attacker uses the ARP responses to obtain the WEP key in less than 6 minutes.

Wireless Intrusion Prevention Concepts

There are three principal ways to secure a wireless network.

- For closed networks (like home users and organizations) the most common way is to configure access restrictions in the access points. Those restrictions may include encryption and checks on MAC address. Wireless Intrusion Prevention Systems can be used to provide wireless LAN security in this network model.

- For commercial providers, hotspots, and large organizations, the preferred solution is often to have an open and unencrypted, but completely isolated wireless network. The users will at first have no access to the Internet nor to any local network resources. Commercial providers usually forward all web traffic to a captive portal which provides for payment and/or authorization. Another solution is to require the users to connect securely to a privileged network using VPN.

- Wireless networks are less secure than wired ones; in many offices intruders can easily visit and hook up their own computer to the wired network without problems, gaining access to the network, and it is also often possible for remote intruders to gain access to the network through backdoors like Back Orifice. One general solution may be end-to-end encryption, with independent authentication on all resources that shouldn't be available to the public.

There is no ready designed system to prevent from fraudulent usage of wireless communication or to protect data and functions with wirelessly communicating computers and other entities. However, there is a system of qualifying the taken measures as a whole according to a common understanding what shall be seen as state of the art. The system of qualifying is an international consensus as specified in ISO/IEC 15408.

A Wireless Intrusion Prevention System

A Wireless Intrusion Prevention System (WIPS) is a concept for the most robust way to counteract wireless security risks. However such WIPS does not exist as a ready designed solution to implement as a software package. A WIPS is typically implemented as an overlay to an existing Wireless LAN infrastructure, although it may be deployed standalone to enforce no-wireless policies within an organization. WIPS is considered so important to wireless security that in July 2009, the Payment Card Industry Security Standards Council published wireless guidelines for PCI DSS recommending the use of WIPS to automate wireless scanning and protection for large organizations.

Security Measures

There are a range of wireless security measures, of varying effectiveness and practicality.

SSID Hiding

A simple but ineffective method to attempt to secure a wireless network is to hide the SSID (Service Set Identifier). This provides very little protection against anything but the most casual intrusion efforts.

MAC ID Filtering

One of the simplest techniques is to only allow access from known, pre-approved MAC addresses. Most wireless access points contain some type of MAC ID filtering. However, an attacker can simply sniff the MAC address of an authorized client and spoof this address.

Static IP Addressing

Typical wireless access points provide IP addresses to clients via DHCP. Requiring clients to set their own addresses makes it more difficult for a casual or unsophisticated intruder to log onto the network, but provides little protection against a sophisticated attacker.

802.11 Security

IEEE 802.1X is the IEEE Standard authentication mechanisms to devices wishing to attach to a Wireless LAN.

Regular WEP

The Wired Equivalent Privacy (WEP) encryption standard was the original encryption standard for wireless, but since 2004 with the ratification WPA2 the IEEE has declared it "deprecated", and while often supported, it is seldom or never the default on modern equipment.

Concerns were raised about its security as early as 2001, dramatically demonstrated in 2005 by the FBI, yet in 2007 T.J. Maxx admitted a massive security breach due in part to a reliance on WEP and the Payment Card Industry took until 2008 to prohibit its use - and even then allowed existing use to continue until June 2010.

WPAv1

The Wi-Fi Protected Access (WPA and WPA2) security protocols were later created to address the problems with WEP. If a weak password, such as a dictionary word or short character string is used, WPA and WPA2 can be cracked. Using a long enough random password (e.g. 14 random letters) or passphrase (e.g. 5 randomly chosen words) makes pre-shared key WPA virtually uncrackable. The second generation of the WPA security protocol (WPA2) is based on the final IEEE 802.11i amendment to the 802.11 standard and is eligible for FIPS 140-2 compliance. With all those encryption schemes, any client in the network that knows the keys can read all the traffic.

Wi-Fi Protected Access (WPA) is a software/firmware improvement over WEP. All regular WLAN-equipment that worked with WEP are able to be simply upgraded and no new equipment needs to be bought. WPA is a trimmed-down version of the 802.11i security standard that was developed by the IEEE 802.11 to replace WEP. The TKIP encryption algorithm was developed for WPA to provide improvements to WEP that could be fielded as firmware upgrades to existing 802.11 devices. The WPA profile also provides optional support for the AES-CCMP algorithm that is the preferred algorithm in 802.11i and WPA2.

WPA Enterprise provides RADIUS based authentication using 802.1x. WPA Personal uses a pre-shared Shared Key (PSK) to establish the security using an 8 to 63 character passphrase. The PSK may also be entered as a 64 character hexadecimal string. Weak PSK passphrases can be broken using off-line dictionary attacks by capturing the messages in the four-way exchange when the client reconnects after being deauthenticated. Wireless suites such as aircrack-ng can crack a weak passphrase in less than a minute. Other WEP/WPA crackers are AirSnort and Auditor Security Collection. Still, WPA Personal is secure when used with 'good' passphrases or a full 64-character hexadecimal key.

There was information, however, that Erik Tews (the man who created the fragmentation attack against WEP) was going to reveal a way of breaking the WPA TKIP implementation at Tokyo's PacSec security conference in November 2008, cracking the encryption on a packet in between 12–15 minutes. Still, the announcement of this 'crack' was somewhat overblown by the media, because as of August, 2009, the best attack on

WPA (the Beck-Tews attack) is only partially successful in that it only works on short data packets, it cannot decipher the WPA key, and it requires very specific WPA implementations in order to work.

Additions to WPAv1

In addition to WPAv1, TKIP, WIDS and EAP may be added alongside. Also, VPN-networks (non-continuous secure network connections) may be set up under the 802.11-standard. VPN implementations include PPTP, L2TP, IPsec and SSH. However, this extra layer of security may also be cracked with tools such as Anger, Deceit and Ettercap for PPTP; and ike-scan, IKEProbe, ipsectrace, and IKEcrack for IPsec-connections.

TKIP

This stands for Temporal Key Integrity Protocol and the acronym is pronounced as tee-kip. This is part of the IEEE 802.11i standard. TKIP implements per-packet key mixing with a re-keying system and also provides a message integrity check. These avoid the problems of WEP.

EAP

The WPA-improvement over the IEEE 802.1X standard already improved the authentication and authorization for access of wireless and wired LANs. In addition to this, extra measures such as the Extensible Authentication Protocol (EAP) have initiated an even greater amount of security. This, as EAP uses a central authentication server. Unfortunately, during 2002 a Maryland professor discovered some shortcomings. Over the next few years these shortcomings were addressed with the use of TLS and other enhancements. This new version of EAP is now called Extended EAP and is available in several versions; these include: EAP-MD5, PEAPv0, PEAPv1, EAP-MSCHAPv2, LEAP, EAP-FAST, EAP-TLS, EAP-TTLS, MSCHAPv2, and EAP-SIM.

EAP-versions

EAP-versions include LEAP, PEAP and other EAP's.

LEAP

This stands for the Lightweight Extensible Authentication Protocol. This protocol is based on 802.1X and helps minimize the original security flaws by using WEP and a sophisticated key management system. This EAP-version is safer than EAP-MD5. This also uses MAC address authentication. LEAP is not secure; THC-LeapCracker can be used to break Cisco's version of LEAP and be used against computers connected to an access point in the form of a dictionary attack. Anwrap and asleap finally are other crackers capable of breaking LEAP.

PEAP

This stands for Protected Extensible Authentication Protocol. This protocol allows for a secure transport of data, passwords, and encryption keys without the need of a certificate server. This was developed by Cisco, Microsoft, and RSA Security.

Other EAPs There are other types of Extensible Authentication Protocol implementations that are based on the EAP framework. The framework that was established supports existing EAP types as well as future authentication methods. EAP-TLS offers very good protection because of its mutual authentication. Both the client and the network are authenticated using certificates and per-session WEP keys. EAP-FAST also offers good protection. EAP-TTLS is another alternative made by Certicom and Funk Software. It is more convenient as one does not need to distribute certificates to users, yet offers slightly less protection than EAP-TLS.

Restricted Access Networks

Solutions include a newer system for authentication, IEEE 802.1x, that promises to enhance security on both wired and wireless networks. Wireless access points that incorporate technologies like these often also have routers built in, thus becoming wireless gateways.

End-to-end Encryption

One can argue that both layer 2 and layer 3 encryption methods are not good enough for protecting valuable data like passwords and personal emails. Those technologies add encryption only to parts of the communication path, still allowing people to spy on the traffic if they have gained access to the wired network somehow. The solution may be encryption and authorization in the application layer, using technologies like SSL, SSH, GnuPG, PGP and similar.

The disadvantage with the end-to-end method is, it may fail to cover all traffic. With encryption on the router level or VPN, a single switch encrypts all traffic, even UDP and DNS lookups. With end-to-end encryption on the other hand, each service to be secured must have its encryption "turned on", and often every connection must also be "turned on" separately. For sending emails, every recipient must support the encryption method, and must exchange keys correctly. For Web, not all web sites offer https, and even if they do, the browser sends out IP addresses in clear text.

The most prized resource is often access to Internet. An office LAN owner seeking to restrict such access will face the nontrivial enforcement task of having each user authenticate themselves for the router.

802.11i Security

The newest and most rigorous security to implement into WLAN's today is the 802.11i

RSN-standard. This full-fledged 802.11i standard (which uses WPAv2) however does require the newest hardware (unlike WPAv1), thus potentially requiring the purchase of new equipment. This new hardware required may be either AES-WRAP (an early version of 802.11i) or the newer and better AES-CCMP-equipment. One should make sure one needs WRAP or CCMP-equipment, as the 2 hardware standards are not compatible.

WPAv2

WPA2 is a WiFi Alliance branded version of the final 802.11i standard. The primary enhancement over WPA is the inclusion of the AES-CCMP algorithm as a mandatory feature. Both WPA and WPA2 support EAP authentication methods using RADIUS servers and preshared key (PSK).

The number of WPA and WPA2 networks are increasing, while the number of WEP networks are decreasing, because of the security vulnerabilities in WEP.

WPA2 has been found to have at least one security vulnerability, nicknamed Hole196. The vulnerability uses the WPA2 Group Temporal Key (GTK), which is a shared key among all users of the same BSSID, to launch attacks on other users of the same BSSID. It is named after page 196 of the IEEE 802.11i specification, where the vulnerability is discussed. In order for this exploit to be performed, the GTK must be known by the attacker.

Additions to WPAv2

Unlike 802.1X, 802.11i already has most other additional security-services such as TKIP. Just as with WPAv1, WPAv2 may work in cooperation with EAP and a WIDS.

WAPI

This stands for WLAN Authentication and Privacy Infrastructure. This is a wireless security standard defined by the Chinese government.

Smart Cards, USB Tokens, and Software Tokens

This is a very strong form of security. When combined with some server software, the hardware or software card or token will use its internal identity code combined with a user entered PIN to create a powerful algorithm that will very frequently generate a new encryption code. The server will be time synced to the card or token. This is a very secure way to conduct wireless transmissions. Companies in this area make USB tokens, software tokens, and smart cards. They even make hardware versions that double as an employee picture badge. Currently the safest security measures are the smart cards / USB tokens. However, these are expensive. The next safest methods are WPA2 or WPA with a RADIUS server. Any one of the three will provide a good base foundation for security. The third item on the list is to educate both employees and contractors on security risks and personal preventive measures. It is also IT's task to keep the company workers' knowledge base up-to-date on any new dangers that they should be cautious about. If the employees are educated, there will be a much lower chance that anyone will accidentally cause a breach in security by not locking down their laptop or bring in a wide open home access point to extend their mobile range. Employees need to be made aware that company laptop security extends to outside of their site walls as well. This includes places such as coffee houses where workers can be at their most vulnerable. The last item on the list deals with 24/7 active defense measures to ensure that the company network is secure and compliant. This can take the form of regularly looking at access point, server, and firewall logs to try to detect any unusual activity. For instance, if any large files went through an access point in the early hours of the morning, a serious investigation into the incident would be called for. There are a number of software and hardware devices that can be used to supplement the usual logs and usual other safety measures.

RF Shielding

It's practical in some cases to apply specialized wall paint and window film to a room or building to significantly attenuate wireless signals, which keeps the signals from propagating outside a facility. This can significantly improve wireless security because it's difficult for hackers to receive the signals beyond the controlled area of an enterprise, such as within parking lots.

Denial of Service Defense

Most DoS attacks are easy to detect. However, a lot of them are difficult to stop even after detection. Here are three of the most common ways to stop a DoS attack.

Black Holing

Black holing is one possible way of stopping a DoS attack. This is a situation where we drop all IP packets from an attacker. This is not a very good long-term strategy because attackers can change their source address very quickly.

This may have negative effects if done automatically. An attacker could knowingly spoof attack packets with the IP address of a corporate partner. Automated defenses could block legitimate traffic from that partner and cause additional problems.

Validating the Handshake

Validating the handshake involves creating false opens, and not setting aside resources until the sender acknowledges. Some firewalls address SYN floods by pre-validating the TCP handshake. This is done by creating false opens. Whenever a SYN segment arrives, the firewall sends back a SYN/ACK segment, without passing the SYN segment on to the target server.

Only when the firewall gets back an ACK, which would happen only in a legitimate connection, would the firewall send the original SYN segment on to the server for which it was originally intended. The firewall doesn't set aside resources for a connection when a SYN segment arrives, so handling a large number of false SYN segments is only a small burden.

Rate Limiting

Rate limiting can be used to reduce a certain type of traffic down to an amount the can be reasonably dealt with. Broadcasting to the internal network could still be used, but only at a limited rate for example. This is for more subtle DoS attacks. This is good if an attack is aimed at a single server because it keeps transmission lines at least partially open for other communication.

Rate limiting frustrates both the attacker, and the legitimate users. This helps but does not fully solve the problem. Once DoS traffic clogs the access line going to the internet, there is nothing a border firewall can do to help the situation. Most DoS attacks are problems of the community which can only be stopped with the help of ISP's and organizations whose computers are taken over as bots and used to attack other firms.

Mobile Devices

With increasing number of mobile devices with 802.1x interfaces, security of such mobile devices becomes a concern. While open standards such as Kismet are targeted towards securing laptops, access points solutions should extend towards covering mobile devices also. Host based solutions for mobile handsets and PDA's with 802.1x interface.

Security within mobile devices fall under three categories:

1. Protecting against ad hoc networks

2. Connecting to rogue access points

3. Mutual authentication schemes such as WPA2 as described above

Wireless IPS solutions now offer wireless security for mobile devices.

Mobile patient monitoring devices are becoming an integral part of healthcare industry and these devices will eventually become the method of choice for accessing and implementing health checks for patients located in remote areas. For these types of patient monitoring systems, security and reliability are critical, because they can influence the condition of patients, and could leave medical professionals in the dark about the condition of the patient if compromised.

Implementing Network Encryption

In order to implement 802.11i, one must first make sure both that the router/access point(s), as well as all client devices are indeed equipped to support the network en-cryption. If this is done, a server such as RADIUS, ADS, NDS, or LDAP needs to be integrated. This server can be a computer on the local network, an access point / router with integrated authentication server, or a remote server. AP's/routers with integrated authentication servers are often very expensive and specifically an option for commer-cial usage like hot spots. Hosted 802.1X servers via the Internet require a monthly fee; running a private server is free yet has the disadvantage that one must set it up and that the server needs to be on continuously.

To set up a server, server and client software must be installed. Server software re-quired is an enterprise authentication server such as RADIUS, ADS, NDS, or LDAP. The required software can be picked from various suppliers as Microsoft, Cisco, Funk Software, Meetinghouse Data, and from some open-source projects. Software includes:

- Aradial RADIUS Server

- Cisco Secure Access Control Software

- freeRADIUS (open-source)

- Funk Software Steel Belted RADIUS (Odyssey)

- Microsoft Internet Authentication Service

- Meetinghouse Data EAGIS

- SkyFriendz (free cloud solution based on freeRADIUS)

Client software comes built-in with Windows XP and may be integrated into other OS's using any of following software:

- AEGIS-client

- Cisco ACU-client

- Intel PROSet/Wireless Software

- Odyssey client

- Xsupplicant (open1X)-project

RADIUS

Remote Authentication Dial In User Service (RADIUS) is an AAA (authentication, au-thorization and accounting) protocol used for remote network access. RADIUS was originally proprietary but was later published under ISOC documents RFC 2138 and RFC 2139. The idea is to have an inside server act as a gatekeeper by verifying identi-ties through a username and password that is already pre-determined by the user. A

RADIUS server can also be configured to enforce user policies and restrictions as well as record accounting information such as connection time for purposes such as billing.

Open Access Points

Today, there is almost full wireless network coverage in many urban areas - the infrastructure for the wireless community network (which some consider to be the future of the internet) is already in place. One could roam around and always be connected to Internet if the nodes were open to the public, but due to security concerns, most nodes are encrypted and the users don't know how to disable encryption. Many people consider it proper etiquette to leave access points open to the public, allowing free access to Internet. Others think the default encryption provides substantial protection at small inconvenience, against dangers of open access that they fear may be substantial even on a home DSL router.

The density of access points can even be a problem - there are a limited number of channels available, and they partly overlap. Each channel can handle multiple networks, but places with many private wireless networks (for example, apartment complexes), the limited number of Wi-Fi radio channels might cause slowness and other problems.

According to the advocates of Open Access Points, it shouldn't involve any significant risks to open up wireless networks for the public:

- The wireless network is after all confined to a small geographical area. A computer connected to the Internet and having improper configurations or other security problems can be exploited by anyone from anywhere in the world, while only clients in a small geographical range can exploit an open wireless access point. Thus the exposure is low with an open wireless access point, and the risks with having an open wireless network are small. However, one should be aware that an open wireless router will give access to the local network, often including access to file shares and printers.

- The only way to keep communication truly secure is to use end-to-end encryption. For example, when accessing an internet bank, one would almost always use strong encryption from the web browser and all the way to the bank - thus it shouldn't be risky to do banking over an unencrypted wireless network. The argument is that anyone can sniff the traffic applies to wired networks too, where system administrators and possible hackers have access to the links and can read the traffic. Also, anyone knowing the keys for an encrypted wireless network can gain access to the data being transferred over the network.

- If services like file shares, access to printers etc. are available on the local net, it is advisable to have authentication (i.e. by password) for accessing it (one should never assume that the private network is not accessible from the outside). Correctly set up, it should be safe to allow access to the local network to outsiders.

- With the most popular encryption algorithms today, a sniffer will usually be able to compute the network key in a few minutes.

- It is very common to pay a fixed monthly fee for the Internet connection, and not for the traffic - thus extra traffic will not be detrimental.

- Where Internet connections are plentiful and cheap, freeloaders will seldom be a prominent nuisance.

On the other hand, in some countries including Germany, persons providing an open access point may be made (partially) liable for any illegal activity conducted via this access point. Also, many contracts with ISPs specify that the connection may not be shared with other persons.

References

- "Health Protection Agency announces further research into use of WiFi". Health Protection Agency. Retrieved 28 August 2008

- Guowang Miao, Jens Zander, Ki Won Sung, and Ben Slimane, Fundamentals of Mobile Data Networks, Cambridge University Press, ISBN 1107143217, 2016

- Alexander Graham Bell (October 1880). "On the Production and Reproduction of Sound by Light". American Journal of Science, Third Series. XX (118): 305–324. also published as "Selenium and the Photophone" in Nature, September 1880

- M. Abolhasan, J. Lipman, W. Ni and B. Hagelstein, "Software-defined wireless networking: centralized, distributed, or hybrid?," in IEEE Network, vol. 29, no. 4, pp. 32-38, July–August 2015. doi:10.1109/MNET.2015.7166188

- Daniels, Nicki (11 December 2006). "Wi-fi: should we be worried?". The Times. London. Retrieved 16 September 2007

- Esptein, Reid J. "Waukesha could be next city to go Wi-Fi" Archived 2006-02-07 at the Wayback Machine. Milwaukee Journal-Sentinel, February 3, 2006

- Joshua Bardwell; Devin Akin (2005). CWNA Official Study Guide (Third ed.). McGraw-Hill. p. 435. ISBN 0-07-225538-2

- Golbon-Haghighi, M.H. (2016). Beamforming in Wireless Networks (PDF). InTech Open. pp. 163–199. ISBN 9781466557529. doi:10.5772/66.399

- Porto, D. C. F.; Cavalcanti, G.; Elias, G. (1 April 2009). "A Layered Routing Architecture for Infrastructure Wireless Mesh Networks". Fifth International Conference on Networking and Services, 2009. ICNS '09: 366–369. doi:10.1109/ICNS.2009.91. Retrieved 14 November 2016

- Trevey, Mick (2007-08-09). "Citywide Wi-Fi Might Not Happen". Local & Regional News. Journal Broadcast Group. Retrieved 2007-08-18

- Mary Kay Carson (2007). Alexander Graham Bell: Giving Voice To The World. Sterling Biographies. New York: Sterling Publishing. pp. 76–78. ISBN 978-1-4027-3230-0

- Ellig, Jerry (November 2006). "A Dynamic Perspective on Government Broadband Initiatives" (PDF). Reason Magazine. Retrieved 2007-08-18

- Pathak, P. H.; Dutta, R. (2011). "A Survey of Network Design Problems and Joint Design Approaches in Wireless Mesh Networks". IEEE Communications Surveys & Tutorials. 13 (3): 396–428. doi:10.1109/SURV.2011.060710.00062

- Dean Tamara (2010). Network+ Guide to Networks (5th ed.). Boston: Cengage Learning. ISBN 978-1-4239-0245-4

- Shaker Alanazi, Kashif Saleem, Jalal Al-Muhtadi, and Abdelouahid Derhab, "Analysis of Denial of Service Impact on Data Routing in Mobile eHealth Wireless Mesh Network," Mobile Information Systems, vol. 2016, Article ID 4853924, 19 pages, 2016. doi:10.1155/2016/48539

An Overview of Wireless LAN

A wireless LAN is used in connecting two or more devices to a wireless network. The types of wireless LANs are infrastructure, peer-to-peer, wireless distribution system and bridge. The chapter strategically encompasses and incorporates the major components and key concepts of wireless LAN, providing a complete understanding.

Wireless LAN

WA wireless local area network (WLAN) is a wireless computer network that links two or more devices using a wireless distribution method (often spread-spectrum or OFDM radio) within a limited area such as a home, school, computer laboratory, or office building. This gives users the ability to move around within a local coverage area and yet still be connected to the network. A WLAN can also provide a connection to the wider Internet.

Most modern WLANs are based on IEEE 802.11 standards and are marketed under the Wi-Fi brand name.

Wireless LANs have become popular for use in the home, due to their ease of installation and use. They are also popular in commercial complexes that offer wireless access to their customers (often without charge).

History

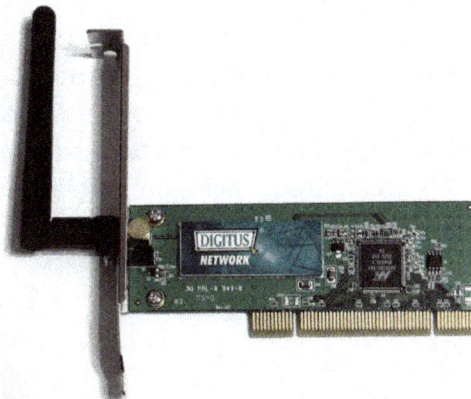

54 Mbit/s WLAN PCI Card (802.11g)

Norman Abramson, a professor at the University of Hawaii, developed the world's first wireless computer communication network, ALOHAnet (operational in 1971), using low-cost ham-like radios. The system included seven computers deployed over four islands to communicate with the central computer on the Oahu Island without using phone lines.

Wireless LAN hardware initially cost so much that it was only used as an alternative to cabled LAN in places where cabling was difficult or impossible. Early development included industry-specific solutions and proprietary protocols, but at the end of the 1990s these were replaced by standards, primarily the various versions of IEEE 802.11 (in products using the Wi-Fi brand name). Beginning in 1991, a European alternative known as HiperLAN/1 was pursued by the European Telecommunications Standards Institute (ETSI) with a first version approved in 1996. This was followed by a HiperLAN/2 functional specification with ATM influences accomplished February 2000. Neither European standard achieved the commercial success of 802.11, although much of the work on HiperLAN/2 has survived in the physical specification (PHY) for IEEE 802.11a, which is nearly identical to the PHY of HiperLAN/2.

In 2009 802.11n was added to 802.11. It operates in both the 2.4 GHz and 5 GHz bands at a maximum data transfer rate of 600 Mbit/s. Most newer routers are able to utilise both wireless bands, known as dualband. This allows data communications to avoid the crowded 2.4 GHz band, which is also shared with Bluetooth devices and microwave ovens. The 5 GHz band is also wider than the 2.4 GHz band, with more channels, which permits a greater number of devices to share the space. Not all channels are available in all regions.

A HomeRF group formed in 1997 to promote a technology aimed for residential use, but it disbanded at the end of 2002.

Architecture

Stations

All components that can connect into a wireless medium in a network are referred to as stations (STA). All stations are equipped with wireless network interface controllers (WNICs). Wireless stations fall into two categories: wireless access points, and clients. Access points (APs), normally wireless routers, are base stations for the wireless network. They transmit and receive radio frequencies for wireless enabled devices to communicate with. Wireless clients can be mobile devices such as laptops, personal digital assistants, IP phones and other smartphones, or non-portable devices such as desktop computers and workstations that are equipped with a wireless network interface.

Basic Service Set

The basic service set (BSS) is a set of all stations that can communicate with each other at PHY layer. Every BSS has an identification (ID) called the BSSID, which is the MAC address of the access point servicing the BSS.

There are two types of BSS: Independent BSS (also referred to as IBSS), and infrastructure BSS. An independent BSS (IBSS) is an ad hoc network that contains no access points, which means they cannot connect to any other basic service set.

Extended Service Set

An extended service set (ESS) is a set of connected BSSs. Access points in an ESS are connected by a distribution system. Each ESS has an ID called the SSID which is a 32-byte (maximum) character string.

Distribution System

A distribution system (DS) connects access points in an extended service set. The concept of a DS can be used to increase network coverage through roaming between cells.

DS can be wired or wireless. Current wireless distribution systems are mostly based on WDS or MESH protocols, though other systems are in use.

Types of Wireless LANs

The IEEE 802.11 has two basic modes of operation: infrastructure and *ad hoc* mode. In *ad hoc* mode, mobile units transmit directly peer-to-peer. In infrastructure mode, mobile units communicate through an access point that serves as a bridge to other networks (such as Internet or LAN).

Since wireless communication uses a more open medium for communication in comparison to wired LANs, the 802.11 designers also included encryption mechanisms: Wired Equivalent Privacy (WEP, now insecure), Wi-Fi Protected Access (WPA, WPA2), to secure wireless computer networks. Many access points will also offer Wi-Fi Protected Setup, a quick (but now insecure) method of joining a new device to an encrypted network.

Infrastructure

Most Wi-Fi networks are deployed in infrastructure mode.

In infrastructure mode, a base station acts as a wireless access point hub, and nodes communicate through the hub. The hub usually, but not always, has a wired or fiber network connection, and may have permanent wireless connections to other nodes.

Wireless access points are usually fixed, and provide service to their client nodes within range.

Wireless clients, such as laptops, smartphones etc. connect to the access point to join the network.

Sometimes a network will have a multiple access points, with the same 'SSID' and security arrangement. In that case connecting to any access point on that network joins the client to the network. In that case, the client software will try to choose the access point to try to give the best service, such as the access point with the strongest signal.

Peer-to-peer

Peer-to-Peer / Ad-Hoc

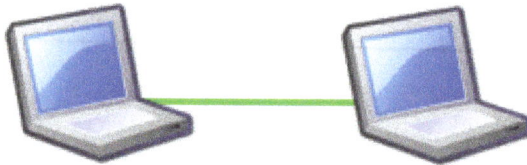

Peer-to-Peer or ad hoc wireless LAN

An ad hoc network (not the same as a WiFi Direct network) is a network where stations communicate only peer to peer (P2P). There is no base and no one gives permission to talk. This is accomplished using the Independent Basic Service Set (IBSS).

A WiFi Direct network is another type of network where stations communicate peer to peer.

In a Wi-Fi P2P group, the group owner operates as an access point and all other devices are clients. There are two main methods to establish a group owner in the Wi-Fi Direct group. In one approach, the user sets up a P2P group owner manually. This method is also known as Autonomous Group Owner (autonomous GO). In the second method, also called negotiation-based group creation, two devices compete based on the group owner intent value. The device with higher intent value becomes a group owner and the second device becomes a client. Group owner intent value can depend on whether the wireless device performs a cross-connection between an infrastructure WLAN service and a P2P group, remaining power in the wireless device, whether the wireless device is already a group owner in another group and/or a received signal strength of the first wireless device.

A peer-to-peer network allows wireless devices to directly communicate with each other. Wireless devices within range of each other can discover and communicate directly without involving central access points. This method is typically used by two computers so that they can connect to each other to form a network. This can basically occur in devices within a closed range.

If a signal strength meter is used in this situation, it may not read the strength accurately and can be misleading, because it registers the strength of the strongest signal, which may be the closest computer.

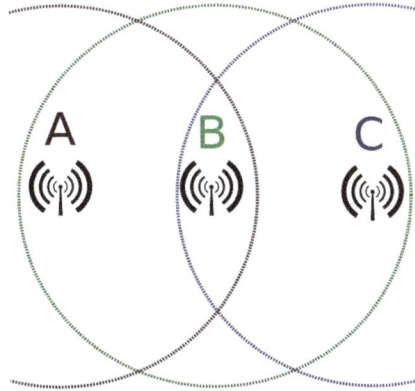

Hidden node problem: Devices A and C are both communicating with B,
but are unaware of each other.

IEEE 802.11 defines the physical layer (PHY) and MAC (Media Access Control) layers based on CSMA/CA (Carrier Sense Multiple Access with Collision Avoidance). The 802.11 specification includes provisions designed to minimize collisions, because two mobile units may both be in range of a common access point, but out of range of each other.

Bridge

A bridge can be used to connect networks, typically of different types. A wireless Ethernet bridge allows the connection of devices on a wired Ethernet network to a wireless network. The bridge acts as the connection point to the Wireless LAN.

Wireless Distribution System

A Wireless Distribution System enables the wireless interconnection of access points in an IEEE 802.11 network. It allows a wireless network to be expanded using multiple access points without the need for a wired backbone to link them, as is traditionally required. The notable advantage of DS over other solutions is that it preserves the MAC addresses of client packets across links between access points.

An access point can be either a main, relay or remote base station. A main base station is typically connected to the wired Ethernet. A relay base station relays data between remote base stations, wireless clients or other relay stations to either a main or another relay base station. A remote base station accepts connections from wireless clients and passes them to relay or main stations. Connections between "clients" are made using MAC addresses rather than by specifying IP assignments.

All base stations in a Wireless Distribution System must be configured to use the same radio channel, and share WEP keys or WPA keys if they are used. They can be configured to different service set identifiers. WDS also requires that every base station be configured to forward to others in the system as mentioned above.

WDS may also be referred to as repeater mode because it appears to bridge and accept wireless clients at the same time (unlike traditional bridging). It should be noted, however, that throughput in this method is halved for all clients connected wirelessly.

When it is difficult to connect all of the access points in a network by wires, it is also possible to put up access points as repeaters.

Roaming

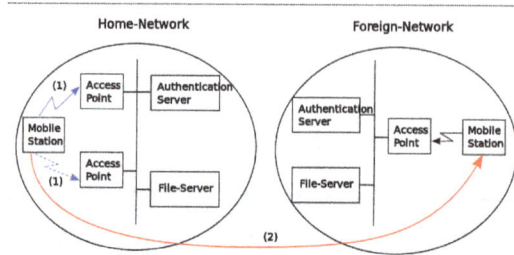

Roaming among Wireless Local Area Networks

There are two definitions for wireless LAN roaming:

1. Internal Roaming: The Mobile Station (MS) moves from one access point (AP) to another AP within a home network if the signal strength is too weak. An authentication server (RADIUS) performs the re-authentication of MS via 802.1x (e.g. with PEAP). The billing of QoS is in the home network. A Mobile Station roaming from one access point to another often interrupts the flow of data among the Mobile Station and an application connected to the network. The Mobile Station, for instance, periodically monitors the presence of alternative access points (ones that will provide a better connection). At some point, based on proprietary mechanisms, the Mobile Station decides to re-associate with an access point having a stronger wireless signal. The Mobile Station, however, may lose a connection with an access point before associating with another access point. In order to provide reliable connections with applications, the Mobile Station must generally include software that provides session persistence.

2. External Roaming: The MS (client) moves into a WLAN of another Wireless Internet Service Provider (WISP) and takes their services (Hotspot). The user can independently of his home network use another foreign network, if this is open for visitors. There must be special authentication and billing systems for mobile services in a foreign network.

Applications

Wireless LANs have a great deal of applications. Modern implementations of WLANs range from small in-home networks to large, campus-sized ones to completely mobile networks on airplanes and trains.

Users can access the Internet from WLAN hotspots in restaurants, hotels, and now with portable devices that connect to 3G or 4G networks. Oftentimes these types of public access points require no registration or password to join the network. Others can be accessed once registration has occurred and/or a fee is paid.

Existing Wireless LAN infrastructures can also be used to work as indoor positioning systems with no modification to the existing hardware.

Performance and Throughput

WLAN, organised in various layer 2 variants (IEEE 802.11), has different characteristics. Across all flavours of 802.11, maximum achievable throughputs are either given based on measurements under ideal conditions or in the layer 2 data rates. This, however, does not apply to typical deployments in which data are being transferred between two endpoints of which at least one is typically connected to a wired infrastructure and the other endpoint is connected to an infrastructure via a wireless link.

This means that typically data frames pass an 802.11 (WLAN) medium and are being converted to 802.3 (Ethernet) or vice versa.

Due to the difference in the frame (header) lengths of these two media, the packet size of an application determines the speed of the data transfer. This means that an application which uses small packets (e.g. VoIP) creates a data flow with a high overhead traffic (e.g. a low goodput).

Other factors which contribute to the overall application data rate are the speed with which the application transmits the packets (i.e. the data rate) and the energy with which the wireless signal is received.

The latter is determined by distance and by the configured output power of the communicating devices.

Same references apply to the attached throughput graphs which show measurements of UDP throughput measurements. Each represents an average (UDP) throughput (the error bars are there, but barely visible due to the small variation) of 25 measurements.

Each is with a specific packet size (small or large) and with a specific data rate (10 kbit/s – 100 Mbit/s). Markers for traffic profiles of common applications are included as well. This text and measurements do not cover packet errors but information about this can be found at above references. The table below shows the maximum achievable (application specific) UDP throughput in the same scenarios (same references again) with various difference WLAN (802.11) flavours. The measurement hosts have been 25 meters apart from each other; loss is again ignored.

Wireless Access Point

In computer networking, a wireless access point (WAP), or more generally just access point (AP), is a networking hardware device that allows a Wi-Fi device to connect to a wired network. The AP usually connects to a router (via a wired network) as a stand-alone device, but it can also be an integral component of the router itself. An AP is differentiated from a hotspot, which is the physical location where Wi-Fi access to a WLAN is available.

Linksys "WAP54G" 802.11g wireless access point

Prior to wireless networks, setting up a computer network in a business, home or school often required running many cables through walls and ceilings in order to deliver network access to all of the network-enabled devices in the building. With the creation of the wireless access point, network users were able to add devices that access the network with few or no cables. An AP normally connects directly to a wired Ethernet connection and the AP then provides wireless connections using radio frequency links for other devices to utilize that wired connection. Most APs support the connection of multiple wireless devices to one wired connection. Modern APs are built to support a standard for sending and receiving data using these radio frequencies. Those standards, and the frequencies they use are defined by the IEEE. Most A

Wireless Access Point Vs. ad Hoc Network

Some people confuse wireless access points with wireless ad hoc networks. An ad hoc network uses a connection between two or more devices without using a wireless access point: the devices communicate directly when in range. An ad hoc network is used in situations such as a quick data exchange or a multiplayer LAN game because setup is easy and does not require an access point. Due to its peer-to-peer layout, ad hoc connections are similar to Bluetooth ones.

But ad hoc connections are generally not recommended for a permanent installation. The reason is that Internet access via ad hoc networks, using features like Windows' Internet Connection Sharing, may work well with a small number of devices that are close to each other, but ad hoc networks don't scale well. Internet traffic will converge to the nodes with direct internet connection, potentially congesting these nodes. For internet-enabled nodes, access points have a clear advantage, with the possibility of having a wired LAN.

Limitations

It is generally recommended that one IEEE 802.11 AP should have, at a maximum, 15-25 clients per radio (most APs having between 1 and 4 radios). However, the actual maximum number of clients that can be supported can vary significantly depending on several factors, such as type of APs in use, density of client environment, desired client throughput, etc. The range of communication can also vary significantly, depending on such variables as indoor or outdoor placement, height above ground, nearby obstructions, other electronic devices that might actively interfere with the signal by broadcasting on the same frequency, type of antenna, the current weather, operating radio frequency, and the power output of devices. Network designers can extend the range of APs through the use of repeaters, which amplify a radio signal, and reflectors, which only bounce it. In experimental conditions, wireless networking has operated over distances of several hundred kilometers.

Most jurisdictions have only a limited number of frequencies legally available for use by wireless networks. Usually, adjacent APs will use different frequencies (Channels) to communicate with their clients in order to avoid interference between the two nearby systems. Wireless devices can "listen" for data traffic on other frequencies, and can rapidly switch from one frequency to another to achieve better reception. However, the limited number of frequencies becomes problematic in crowded downtown areas with tall buildings using multiple APs. In such an environment, signal overlap becomes an issue causing interference, which results in signal droppage and data errors.

Wireless networking lags wired networking in terms of increasing bandwidth and throughput. While (as of 2013) high-density 256-QAM (TurboQAM) modulation, 3-antenna wireless devices for the consumer market can reach sustained real-world speeds of some 240 Mbit/s at 13 m behind two standing walls (NLOS) depending on their nature or 360 Mbit/s at 10 m line of sight or 380 Mbit/s at 2 m line of sight (IEEE 802.11ac) or 20 to 25 Mbit/s at 2 m line of sight (IEEE 802.11g), wired hardware of similar cost reaches closer to 1000 Mbit/s up to specified distance of 100 m with twisted-pair cabling in optimal conditions (Category 5 (known as Cat-5) or better cabling with Gigabit Ethernet). One impediment to increasing the speed of wireless communications comes from Wi-Fi's use of a shared communications medium: Thus, two stations in infrastructure mode that are communicating with each other even over the same AP must have each and every frame transmitted twice: from the sender to the AP, then from the AP to the receiver. This approximately halves the effective bandwidth, so an AP is only able to use somewhat less than half the actual over-the-air rate for data throughput. Thus a typical 54 Mbit/s wireless connection actually carries TCP/IP data at 20 to 25 Mbit/s. Users of legacy wired networks expect faster speeds, and people using wireless connections keenly want to see the wireless networks catch up.

By 2012, 802.11n based access points and client devices have already taken a fair share of the marketplace and with the finalization of the 802.11n standard in 2009 inherent problems integrating products from different vendors are less prevalent.

Security

Wireless access has special security considerations. Many wired networks base the security on physical access control, trusting all the users on the local network, but if wireless access points are connected to the network, anybody within range of the AP (which typically extends farther than the intended area) can attach to the network.

The most common solution is wireless traffic encryption. Modern access points come with built-in encryption. The first generation encryption scheme, wired equivalent privacy (WEP), proved easy to crack; the second and third generation schemes, WPA and WPA2, are considered secure if a strong enough password or passphrase is used.

Some APs support hotspot style authentication using RADIUS and other authentication servers.

Opinions about wireless network security vary widely. For example, in a 2008 article for *Wired* magazine, Bruce Schneier asserted the net benefits of open Wi-Fi without passwords outweigh the risks, a position supported in 2014 by Peter Eckersley of the Electronic Frontier Foundation.

The opposite position was taken by Nick Mediati in an article for *PC World*, in which he takes the position that every wireless access point should be protected with a password.

Lightweight Access Point Protocol

Lightweight Access Point Protocol or LWAPP is the name of a protocol that can control multiple Wi-Fi wireless access points at once. This can reduce the amount of time spent on configuring, monitoring or troubleshooting a large network. The system will also allow network administrators to closely analyze the network.

This system is installed in a central server that gathers data from RF devices from different brands and settings. The server can command a selected group of devices to apply given settings simultaneously.

Standardization

LWAPP was proposed by Airespace, as a standard protocol to provide interoperability among any brand of access point. Airespace was purchased by Cisco Systems. Its purpose was to standardize "lightweight" access points with the Internet Engineering Task Force (IETF), but it was approved as a standard. Sponsored by Cisco Systems, it has been submitted to IETF in RFC 5412.

Although this protocol has so far not been popular beyond the Airespace/Cisco product lines, the CAPWAP standard is based on LWAPP. Support for LWAPP is also found in analysis products from AirMagnet, who has recently implemented a software based on this protocol to analyze Cisco wireless products.

Still considered proprietary, LWAPP systems compete with other non-standard lightweight wireless mechanisms from companies like Meru Networks and Aruba Networks.

LWAPP Layer 2

On Layer 2, LWAPP only requires a data link connection in order to transfer frames and Layer 2 broadcasts. Even if IP connectivity is not established it will still operate at layer 2.

LWAPP Layer 3 and 4

Layer 4 UDP 12222 (data channel) and 12223 (control channel) connectivity must be established to work with this form of the protocol. Broadcasts or DHCP option 43 can be used to prime the access-points of the network. The controller must be on the same subnet if DHCP is not configured to handle layer 3 LWAPP provisioning. Another option for directing an AP to the controller is by defining the controller on the DNS server of the network.

Wireless Intrusion Prevention System

In computing, a wireless intrusion prevention system (WIPS) is a network device that monitors the radio spectrum for the presence of unauthorized access points *(intrusion detection)*, and can automatically take countermeasures *(intrusion prevention)*.

Purpose

The primary purpose of a WIPS is to prevent unauthorized network access to local area networks and other information assets by wireless devices. These systems are typically implemented as an overlay to an existing Wireless LAN infrastructure, although they may be deployed standalone to enforce no-wireless policies within an organization. Some advanced wireless infrastructure has integrated WIPS capabilities.

Large organizations with many employees are particularly vulnerable to security breaches caused by rogue access points. If an employee (trusted entity) in a location brings in an easily available wireless router, the entire network can be exposed to any-one within range of the signals.

In July 2009, the PCI Security Standards Council published wireless guidelines for PCI DSS recommending the use of WIPS to automate wireless scanning for large organizations.

Intrusion Detection

A wireless intrusion detection system (WIDS) monitors the radio spectrum for the presence of unauthorized, rogue access points and the use of wireless attack tools. The system monitors the radio spectrum used by wireless LANs, and immediately alerts a systems administrator whenever a rogue access point is detected. Conventionally it is achieved by comparing the MAC address of the participating wireless devices.

Rogue devices can spoof MAC address of an authorized network device as their own. New research uses fingerprinting approach to weed out devices with spoofed MAC addresses. The idea is to compare the unique signatures exhibited by the signals emitted by each wireless device against the known signatures of pre-authorized, known wireless devices.

Intrusion Prevention

In addition to intrusion detection, a WIPS also includes features that prevent against the threat *automatically*. For automatic prevention, it is required that the WIPS is able to accurately detect and automatically classify a threat.

The following types of threats can be prevented by a good WIPS:

- Rogue access points – WIPS should understand the difference between rogue APs and external (neighbor's) APs

- Mis-configured AP

- Client mis-association

- Unauthorized association

- Man-in-the-middle attack

- *Ad hoc* networks

- MAC spoofing

- Honeypot / evil twin attack

- Denial-of-service attack

Implementation

WIPS configurations consist of three components:

- Sensors — These devices contain antennas and radios that scan the wireless spectrum for packets and are installed throughout areas to be protected

- Server — The WIPS server centrally analyzes packets captured by sensors

- Console — The console provides the primary user interface into the system for administration and reporting

A simple intrusion detection system can be a single computer, connected to a wireless signal processing device, and antennas placed throughout the facility. For huge organizations, a Multi Network Controller provides central control of multiple WIPS servers, while for SOHO or SMB customers, all the functionality of WIPS is available in single box.

In a WIPS implementation, users first define the operating wireless policies in the WIPS. The WIPS sensors then analyze the traffic in the air and send this information to WIPS server. The WIPS server correlates the information, validates it against the defined policies, and classifies if it is a threat. The administrator of the WIPS is then notified of the threat, or, if a policy has been set accordingly, the WIPS takes automatic protection measures.

WIPS is configured as either a network implementation or a hosted implementation.

Network Implementation

In a network WIPS implementation, server, sensors and the console are all placed inside a private network and are not accessible from the Internet.

Sensors communicate with the server over a private network using a private port. Since the server resides on the private network, users can access the console only from within the private network.

A network implementation is suitable for organizations where all locations are within the private network.

Hosted Implementation

In a hosted WIPS implementation, sensors are installed inside a private network. However, the server is hosted in secure data center and is accessible on the Internet. Users can access the WIPS console from anywhere on the Internet. A hosted WIPS implementation is as secure as a network implementation because the data flow is encrypted between sensors and server, as well as between server and console. A hosted WIPS implementation requires very little configuration because the sensors are programmed to automatically look for the server on the Internet over a secure TLS connection.

For a large organization with locations that are not a part of a private network, a hosted WIPS implementation simplifies deployment significantly because sensors connect to the Server over the Internet without requiring any special configuration. Additionally, the Console can be accessed securely from anywhere on the Internet.

Hosted WIPS implementations are available in an on-demand, subscription-based software as a service model. Hosted implementations may be appropriate for organizations looking to fulfill the minimum scanning requirements of PCI DSS.

WLAN Authentication and Privacy Infrastructure

WLAN Authentication and Privacy Infrastructure (WAPI) is a Chinese National Standard for Wireless LANs (GB 15629.11-2003). Although it was allegedly designed to operate on top of WiFi, compatibility with the security protocol used by the 802.11 wireless networking standard developed by the IEEE is in dispute. Due to the limited access of the standard (only eleven Chinese companies had access), it was the focus of a U.S.-China trade dispute. Following this it was submitted to, and rejected by the ISO. It was resubmitted to ISO in 2010, but was cancelled as a project on 21 November 2011 after being withdrawn by China. Part of the reason for withdrawal is thought to be the well documented observations by IEEE representatives that showed WAPI was equivalent to a small subset of IEEE 802.11i based systems.

How the Standard Works

WAPI, which was initiated to resolve the existing security loopholes (WEP) in WLAN international standard (ISO/IEC 8802-11), was issued to be Chinese national standard in 2003. WAPI works by having a central Authentication Service Unit (ASU) which is known to both the wireless user and the access point and which acts as a central authority verifying both. The WAPI standard (draft JTC1/SC6/N14619) allows selection of the symmetric encryption algorithm, either AES or SMS4, which has been declassified in January 2006 and passed evaluation by independent experts.

Criticism

One argument was WAPI standard used security through obscurity, another is that it was designed to limit trade into China, as well as requiring foreign companies to provide confidential trade secrets to Chinese corporations.

History

US–China Trade Dispute

In late 2003, the Chinese government announced a policy requiring that wireless devices sold in China include WAPI support and foreign companies wanting access to the Chinese market could produce WAPI-compliant products independently or partner with one of 11 Chinese firms to which the standard was disclosed. This issue became a point of trade discussions between the then United States Secretary of State Colin Powell and his Chinese government equivalent. China agreed to indefinitely postpone implementation of the policy.

ISO Rejection

The Chinese Standards Association (SAC: Standardization Administration of the People's Republic of China) subsequently submitted WAPI to the ISO standards organization for recognition as an international standard, at about the same time as the IEEE 802.11i standard. After much debate related to both process issues and technical issues, the IEC/ISO Secretaries General decided to send the proposals to parallel fast track ballots. In March 2006, the 802.11i proposal was approved and the WAPI proposal was rejected. This result was confirmed at a Ballot Resolution meeting held in June 2006, during which the SAC delegation walked out.

The result was subject to two appeals by SAC to the ISO/IEC Secretaries General that alleged "unethical" and "amoral" behavior during the balloting process and irregularities during the ballot resolution process. The official Chinese news agency Xinhua said on May 29, 2006, that appeals were filed in April and May 2006 and, the agency said, alleged that the IEEE was involved in "organizing a conspiracy against the China-developed WAPI, insulting China and other national bodies, and intimidation and threats." Xinhua did not make these allegations specific. In July 2006, 802.11i was published as an ISO/IEC standard. WAPI is no longer being considered by ISO/IEC and all appeals have been dismissed.

After the preliminary results were announced in March 2006, various press reports from China suggested that WAPI may still be mandated in China. TBT (Technical Barrier to Trade) declarations to the WTO in January 2006 and a statement in June 2006 to ISO/IEC JTC1/SC6, in which SAC said they would not respect the status of 802.11i as an international standard, seemed to support this possibility. However, as of early 2007, the only official Chinese policy related to WAPI is a "government preference" for WAPI in government and government-funded systems. It is unclear how strongly this preference has been enforced, and it seems to have had little effect on the non-government market.

ISO Resubmission

In 2009, the China NB was encouraged by SC6 to resubmit WAPI to SC6. It was allocated the standard number ISO/IEC 20011 after passing the first stage of balloting. Positive votes and commitments to participate in the standardisation process were received from China, Korea, Czech Republic, Switzerland and Kenya. Negative votes were received from the US and the UK. The US and the IEEE 802.11 Working Group provided numerous detailed comments rebutting the case for standardisation made by the China NB in the New Project proposal.

The required comment resolution on the ballot only started in June 2011, with the US, UK, China, Korea and Switzerland NBs and the IEEE 802.11 Working Group all participating. The Swiss NB representative admitted during the process that he was a paid consultant to IWNCOMM, the Chinese source of the WAPI technology. The Kenya and Czech NBs did not participate in the comment resolution process or in any other discussions related to WAPI after the close of the ballot in early 2010.

The comment resolution process failed after agreement could not be established on a variety of fundamental issues. For example, the China NB continued to insist that WAPI was justified because 802.11 included WEP, which is known to be broken. On the other side, the US NB and the IEEE 802.11 NB noted that WEP-based security had been deprecated in favour of WPA2-based security in IEEE 802.11-2007, and that no one had ever alleged any issues with WPA2-based security. In addition, the IEEE 802.11 WG noted that the functionality offered by WAPI systems was equivalent to only a small subset of the security offered by WPA2-based systems.

The China NB eventually withdrew WAPI in October 2011 (document JTC1/SC6N15030) and the project formally cancelled by SC6 in February 2012. The reasons for the withdrawal are unclear. The Chinese proponents of WAPI from IWNCOMM were clearly very unhappy when the withdrawal was announced. It has been speculated that Chinese government authorities ordered the withdrawal on the basis that WAPI had failed to be standardised by ISO/IEC after eight years. In addition, despite mandates for WAPI to be implemented in China in Wi-Fi enabled mobile phones and by the three Chinese service providers, it is very rarely used in practice.

Chinese Cell Phone Usage

Mobile phones in China are controlled by MIIT. Mobile phones coming out in China in 2009 required to support the WAPI standard. One of the sticking points behind the iPhone in China was the support of WiFi without the WAPI standard. In the end, it was released without any WLAN at all.

According to China's State Radio Monitoring Center Chinese, in April 2011 regulators approved the frequency ranges used by a new Apple mobile phone with 3G and wireless LAN support including WAPI. Dell Inc's Mini 3 phones have also received network access licenses for China.

The Chinese government's preference for the WAPI standard in some respects is similar to their preference for the TD-SCDMA for their 3G network.

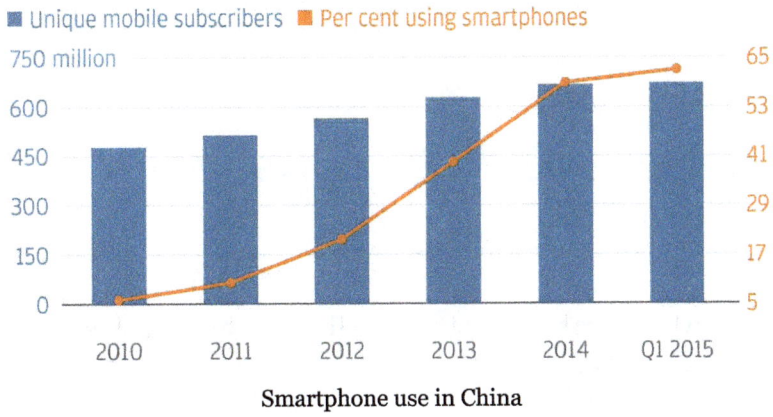

Smartphone use in China

4

Wireless Sensor Networks: An Integrated Study

Wireless sensor networks are mainly built of nodes. The applications can be categorized into area monitoring, environmental sensing, health care monitoring and industrial monitoring. Sensor node, mobile wireless sensor network and optical wireless communications have been explained in the chapter. WiFi technology is best understood in confluence with the major topics listed in the following chapter.

Wireless Sensor Network

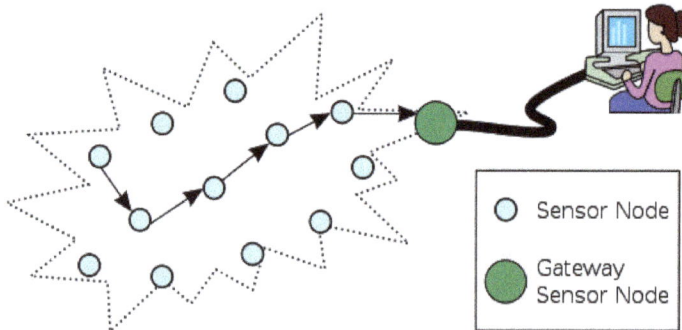

Typical multi-hop wireless sensor network architecture

Wireless sensor networks (WSN), sometimes called wireless sensor and actuator networks (WSAN), are spatially distributed autonomous sensors to *monitor* physical or environmental conditions, such as temperature, sound, pressure, etc. and to cooperatively pass their data through the network to a main location. The more modern networks are bi-directional, also enabling *control* of sensor activity. The development of wireless sensor networks was motivated by military applications such as battlefield surveillance; today such networks are used in many industrial and consumer applications, such as industrial process monitoring and control, machine health monitoring, and so on.

The WSN is built of "nodes" – from a few to several hundreds or even thousands, where each node is connected to one (or sometimes several) sensors. Each such sensor network node has typically several parts: a radio transceiver with an internal antenna or connection to an external antenna, a microcontroller, an electronic circuit for interfacing with the sensors and an energy source, usually a battery or an embedded form of energy harvesting. A sensor node might vary in size from that of a shoebox down to the size of a grain of dust, although functioning "motes" of genuine microscopic dimensions have yet to be created. The cost of sensor nodes is similarly variable, ranging from a few to hundreds of dollars, depending on the complexity of the individual sensor nodes. Size and cost constraints on sensor nodes result in corresponding constraints on resources such as energy, memory, computational speed and communications bandwidth. The topology of the WSNs can vary from a simple star network to an advanced multi-hop wireless mesh network. The propagation technique between the hops of the network can be routing or flooding.

In computer science and telecommunications, wireless sensor networks are an active research area with numerous workshops and conferences arranged each year, for example IPSN, SenSys, and EWSN.

Application

Area Monitoring

Area monitoring is a common application of WSNs. In area monitoring, the WSN is deployed over a region where some phenomenon is to be monitored. A military example is the use of sensors detect enemy intrusion; a civilian example is the geo-fencing of gas or oil pipelines.

Health Care Monitoring

The sensor networks for medical applications can be of several types: implanted, wearable, and environment-embedded. The implantable medical devices are those that are inserted inside human body. Wearable devices are used on the body surface of a human or just at close proximity of the user. Environment-embedded systems employ sensors contained in the environment. Possible applications include body position measurement, location of persons, overall monitoring of ill patients in hospitals and at homes. Devices embedded in the environment track the physical state of a person for continuous health diagnosis, using as input the data from a network of depth cameras, a sensing floor, or other similar devices. Body-area networks can collect information about an individual's health, fitness, and energy expenditure. In health care applications the privacy and authenticity of user data has prime importance. Especially due to the integration of sensor networks, with IoT, the authentication of user become more challenging; however, a solution is presented in recent work.

Environmental/Earth Sensing

There are many applications in monitoring environmental parameters, examples of which are given below. They share the extra challenges of harsh environments and reduced power supply.

Air Pollution Monitoring

Wireless sensor networks have been deployed in several cities (Stockholm, London, and Brisbane) to monitor the concentration of dangerous gases for citizens. These can take advantage of the ad hoc wireless links rather than wired installations, which also make them more mobile for testing readings in different areas.

Forest Fire Detection

A network of Sensor Nodes can be installed in a forest to detect when a fire has started. The nodes can be equipped with sensors to measure temperature, humidity and gases which are produced by fire in the trees or vegetation. The early detection is crucial for a successful action of the firefighters; thanks to Wireless Sensor Networks, the fire brigade will be able to know when a fire is started and how it is spreading.

Landslide Detection

A landslide detection system makes use of a wireless sensor network to detect the slight movements of soil and changes in various parameters that may occur before or during a landslide. Through the data gathered it may be possible to know the impending occurrence of landslides long before it actually happens.

Water Quality Monitoring

Water quality monitoring involves analyzing water properties in dams, rivers, lakes and oceans, as well as underground water reserves. The use of many wireless distributed sensors enables the creation of a more accurate map of the water status, and allows the permanent deployment of monitoring stations in locations of difficult access, without the need of manual data retrieval.

Natural Disaster Prevention

Wireless sensor networks can effectively act to prevent the consequences of natural disasters, like floods. Wireless nodes have successfully been deployed in rivers where changes of the water levels have to be monitored in real time.

Industrial Monitoring

Machine Health Monitoring

Wireless sensor networks have been developed for machinery condition-based maintenance (CBM) as they offer significant cost savings and enable new functionality.

Wireless sensors can be placed in locations difficult or impossible to reach with a wired system, such as rotating machinery and untethered vehicles.

Data Center Monitoring

Due to the high density of servers racks in a data center, often cabling and IP addresses are an issue. To overcome that problem more and more racks are fitted out with wireless temperature sensors to monitor the intake and outtake temperatures of racks. As ASHRAE recommends up to 6 temperature sensors per rack, meshed wireless temperature technology gives an advantage compared to traditional cabled sensors.

Data Logging

Wireless sensor networks are also used for the collection of data for monitoring of environmental information, this can be as simple as the monitoring of the temperature in a fridge to the level of water in overflow tanks in nuclear power plants. The statistical information can then be used to show how systems have been working. The advantage of WSNs over conventional loggers is the "live" data feed that is possible.

Water/Waste Water Monitoring

Monitoring the quality and level of water includes many activities such as checking the quality of underground or surface water and ensuring a country's water infrastructure for the benefit of both human and animal. It may be used to protect the wastage of water.

Structural Health Monitoring

Wireless sensor networks can be used to monitor the condition of civil infrastructure and related geo-physical processes close to real time, and over long periods through data logging, using appropriately interfaced sensors.

Wine Production

Wireless sensor networks are used to monitor wine production, both in the field and the cellar.

Characteristics

The main characteristics of a WSN include:

- Power consumption constraints for nodes using batteries or energy harvesting
- Ability to cope with node failures (resilience)
- Some mobility of nodes
- Heterogeneity of nodes

- Scalability to large scale of deployment

- Ability to withstand harsh environmental conditions

- Ease of use

- Cross-layer design

Cross-layer is becoming an important studying area for wireless communications. In addition, the traditional layered approach presents three main problems:

1. Traditional layered approach cannot share different information among different layers, which leads to each layer not having complete information. The traditional layered approach cannot guarantee the optimization of the entire network.

2. The traditional layered approach does not have the ability to adapt to the environmental change.

3. Because of the interference between the different users, access conflicts, fading, and the change of environment in the wireless sensor networks, traditional layered approach for wired networks is not applicable to wireless networks.

So the cross-layer can be used to make the optimal modulation to improve the transmission performance, such as data rate, energy efficiency, QoS (Quality of Service), etc. Sensor nodes can be imagined as small computers which are extremely basic in terms of their interfaces and their components. They usually consist of a *processing unit* with limited computational power and limited memory, *sensors* or MEMS (including specific conditioning circuitry), a *communication device* (usually radio transceivers or alternatively optical), and a power source usually in the form of a battery. Other possible inclusions are energy harvesting modules, secondary ASICs, and possibly secondary communication interface (e.g. RS-232 or USB).

The base stations are one or more components of the WSN with much more computational, energy and communication resources. They act as a gateway between sensor nodes and the end user as they typically forward data from the WSN on to a server. Other special components in routing based networks are routers, designed to compute, calculate and distribute the routing tables.

Platforms

Hardware

One major challenge in a WSN is to produce *low cost* and *tiny* sensor nodes. There are an increasing number of small companies producing WSN hardware and the commercial situation can be compared to home computing in the 1970s. Many of the nodes are still in the research and development stage, particularly their software. Also inherent to sensor network adoption is the use of very low power methods for radio communication and data acquisition.

In many applications, a WSN communicates with a Local Area Network or Wide Area Network through a gateway. The Gateway acts as a bridge between the WSN and the other network. This enables data to be stored and processed by devices with more resources, for example, in a remotely located server. A wireless wide area network used primarily for low-power devices is known as a Low-Power Wide-Area Network (LP-WAN).

Software

Energy is the scarcest resource of WSN nodes, and it determines the lifetime of WSNs. WSNs may be deployed in large numbers in various environments, including remote and hostile regions, where ad hoc communications are a key component. For this reason, algorithms and protocols need to address the following issues:

- Increased lifespan

- Robustness and fault tolerance

- Self-configuration

Lifetime maximization: Energy/Power Consumption of the sensing device should be minimized and sensor nodes should be energy efficient since their limited energy resource determines their lifetime. To conserve power, wireless sensor nodes normally power off both the radio transmitter and the radio receiver when not in use.

Operating Systems

Operating systems for wireless sensor network nodes are typically less complex than general-purpose operating systems. They more strongly resemble embedded systems, for two reasons. First, wireless sensor networks are typically deployed with a particular application in mind, rather than as a general platform. Second, a need for low costs and low power leads most wireless sensor nodes to have low-power microcontrollers ensuring that mechanisms such as virtual memory are either unnecessary or too expensive to implement.

It is therefore possible to use embedded operating systems such as eCos or uC/OS for sensor networks. However, such operating systems are often designed with real-time properties.

TinyOS is perhaps the first operating system specifically designed for wireless sensor networks. TinyOS is based on an event-driven programming model instead of multithreading. TinyOS programs are composed of *event handlers* and *tasks* with run-to-completion semantics. When an external event occurs, such as an incoming data packet or a sensor reading, TinyOS signals the appropriate event handler to handle the event. Event handlers can post tasks that are scheduled by the TinyOS kernel some time later.

LiteOS is a newly developed OS for wireless sensor networks, which provides UNIX-like abstraction and support for the C programming language.

Contiki is an OS which uses a simpler programming style in C while providing advances such as 6LoWPAN and Protothreads.

Online Collaborative Sensor Data Management Platforms

Online collaborative sensor data management platforms are on-line database services that allow sensor owners to register and connect their devices to feed data into an on-line database for storage and also allow developers to connect to the database and build their own applications based on that data. Examples include Xively and the Wikisensing platform. Such platforms simplify online collaboration between users over diverse data sets ranging from energy and environment data to that collected from transport services. Other services include allowing developers to embed real-time graphs & widgets in websites; analyse and process historical data pulled from the data feeds; send real-time alerts from any datastream to control scripts, devices and environments.

The architecture of the Wikisensing system describes the key components of such systems to include APIs and interfaces for online collaborators, a middleware containing the business logic needed for the sensor data management and processing and a storage model suitable for the efficient storage and retrieval of large volumes of data. 0525162632.

Simulation of WSNs

At present, agent-based modeling and simulation is the only paradigm which allows the simulation of complex behavior in the environments of wireless sensors (such as flocking). Agent-based simulation of wireless sensor and ad hoc networks is a relatively new paradigm. Agent-based modelling was originally based on social simulation.

Network simulators like OPNET, NetSim and NS2 can be used to simulate a wireless sensor network.

Other Concepts

Distributed Sensor Network

If a centralized architecture is used in a sensor network and the central node fails, then the entire network will collapse, however the reliability of the sensor network can be increased by using a distributed control architecture. Distributed control is used in WSNs for the following reasons:

1. Sensor nodes are prone to failure,

2. For better collection of data,

3. To provide nodes with backup in case of failure of the central node.

There is also no centralised body to allocate the resources and they have to be self organized.

Data Integration and Sensor Web

The data gathered from wireless sensor networks is usually saved in the form of numerical data in a central base station. Additionally, the Open Geospatial Consortium (OGC) is specifying standards for interoperability interfaces and metadata encodings that enable real time integration of heterogeneous sensor webs into the Internet, allowing any individual to monitor or control wireless sensor networks through a web browser.

In-network Processing

To reduce communication costs some algorithms remove or reduce nodes' redundant sensor information and avoid forwarding data that is of no use. As nodes can inspect the data they forward, they can measure averages or directionality for example of readings from other nodes. For example, in sensing and monitoring applications, it is generally the case that neighboring sensor nodes monitoring an environmental feature typically register similar values. This kind of data redundancy due to the spatial correlation between sensor observations inspires techniques for in-network data aggregation and mining. Aggregation reduces the amount of network traffic which helps to reduce energy consumption on sensor nodes. Recently, it has been found that network gateways also play an important role in improving energy efficiency of sensor nodes by scheduling more resources for the nodes with more critical energy efficiency need and advanced energy efficient scheduling algorithms need to be implemented at network gateways for the improvement of the overall network energy efficiency.

Secure Data Aggregation

This is a form of in-network processing where sensor nodes are assumed to be unsecured with limited available energy, while the base station is assumed to be secure with unlimited available energy. Aggregation complicates the already existing security challenges for wireless sensor networks and requires new security techniques tailored specifically for this scenario. Providing security to aggregate data in wireless sensor networks is known as *secure data aggregation in WSN*. were the first few works discussing techniques for secure data aggregation in wireless sensor networks.

Two main security challenges in secure data aggregation are confidentiality and integrity of data. While encryption is traditionally used to provide end to end confidentiality in wireless sensor network, the aggregators in a secure data aggregation scenario need to decrypt the encrypted data to perform aggregation. This exposes the plaintext at the aggregators, making the data vulnerable to attacks from an adversary. Similarly an aggregator can inject false data into the aggregate and make the base station accept false data. Thus, while data aggregation improves energy efficiency of a network, it complicates the existing security challenges.

Sensor Node

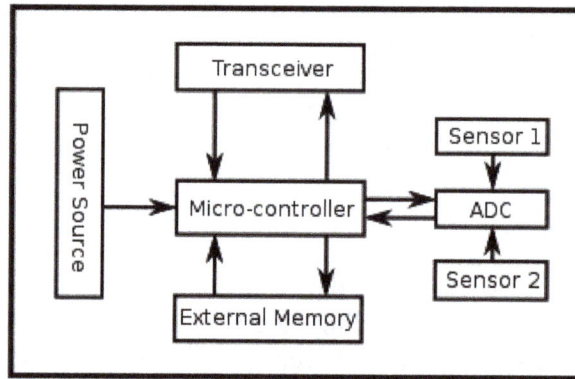

The typical architecture of the sensor node

A sensor node, also known as a mote (chiefly in North America), is a node in a sensor network that is capable of performing some processing, gathering sensory information and communicating with other connected nodes in the network. A mote is a node but a node is not always a mote.

History

Although wireless sensor nodes have existed for decades and used for applications as diverse as earthquake measurements to warfare, the modern development of small sensor nodes dates back to the 1998 Smartdust project and the NASA Sensor Webs Project One of the objectives of the Smartdust project was to create autonomous sensing and communication within a cubic millimeter of space. Though this project ended early on, it led to many more research projects. They include major research centres in Berkeley NEST and CENS. The researchers involved in these projects coined the term *mote* to refer to a sensor node. The equivalent term in the NASA Sensor Webs Project for a physical sensor node is *pod*, although the sensor node in a Sensor Web can be another Sensor Web itself. Physical sensor nodes have been able to increase their capability in conjunction with Moore's Law. The chip footprint contains more complex and lower powered microcontrollers. Thus, for the same node footprint, more silicon capability can be packed into it. Nowadays, motes focus on providing the longest wireless range (dozens of km), the lowest energy consumption (a few uA) and the easiest development process for the user.

Components

The main components of a sensor node are a microcontroller, transceiver, external memory, power source and one or more sensors.

Controller

The controller performs tasks, processes data and controls the functionality of other components in the sensor node. While the most common controller is a microcontroller, other alternatives that can be used as a controller are: a general purpose desktop microprocessor, digital signal processors, FPGAs and ASICs. A microcontroller is often used in many embedded systems such as sensor nodes because of its low cost, flexibility to connect to other devices, ease of programming, and low power consumption. A general purpose microprocessor generally has a higher power consumption than a microcontroller, therefore it is often not considered a suitable choice for a sensor node. Digital Signal Processors may be chosen for broadband wireless communication applications, but in Wireless Sensor Networks the wireless communication is often modest: i.e., simpler, easier to process modulation and the signal processing tasks of actual sensing of data is less complicated. Therefore, the advantages of DSPs are not usually of much importance to wireless sensor nodes. FPGAs can be reprogrammed and reconfigured according to requirements, but this takes more time and energy than desired.

Transceiver

Sensor nodes often make use of ISM band, which gives free radio, spectrum allocation and global availability. The possible choices of wireless transmission media are radio frequency (RF), optical communication (laser) and infrared. Lasers require less energy, but need line-of-sight for communication and are sensitive to atmospheric conditions. Infrared, like lasers, needs no antenna but it is limited in its broadcasting capacity. Radio frequency-based communication is the most relevant that fits most of the WSN applications. WSNs tend to use license-free communication frequencies: 173, 433, 868, and 915 MHz; and 2.4 GHz. The functionality of both transmitter and receiver are combined into a single device known as a transceiver. Transceivers often lack unique identifiers. The operational states are transmit, receive, idle, and sleep. Current generation transceivers have built-in state machines that perform some operations automatically.

Most transceivers operating in idle mode have a power consumption almost equal to the power consumed in receive mode. Thus, it is better to completely shut down the transceiver rather than leave it in the idle mode when it is not transmitting or receiving.

A significant amount of power is consumed when switching from sleep mode to transmit mode in order to transmit a packet.

External Memory

From an energy perspective, the most relevant kinds of memory are the on-chip memory of a microcontroller and Flash memory—off-chip RAM is rarely, if ever, used. Flash memories are used due to their cost and storage capacity. Memory requirements are very much application dependent. Two categories of memory based on the purpose of storage are: user memory used for storing application related or personal data, and program memory used for programming the device. Program memory also contains identification data of the device if present.

Power Source

A wireless sensor node is a popular solution when it is difficult or impossible to run a mains supply to the sensor node. However, since the wireless sensor node is often placed in a hard-to-reach location, changing the battery regularly can be costly and inconvenient. An important aspect in the development of a wireless sensor node is ensuring that there is always adequate energy available to power the system. The sensor node consumes power for sensing, communicating and data processing. More energy is required for data communication than any other process. The energy cost of transmitting 1 Kb a distance of 100 metres (330 ft) is approximately the same as that used for the execution of 3 million instructions by a 100 million instructions per second/W processor. Power is stored either in batteries or capacitors. Batteries, both rechargeable and non-rechargeable, are the main source of power supply for sensor nodes. They are also classified according to electrochemical material used for the electrodes such as NiCd (nickel-cadmium), NiZn (nickel-zinc), NiMH (nickel-metal hydride), and lithium-ion. Current sensors are able to renew their energy from solar sources, temperature differences, or vibration. Two power saving policies used are Dynamic Power Management (DPM) and Dynamic Voltage Scaling (DVS). DPM conserves power by shutting down parts of the sensor node which are not currently used or active. A DVS scheme varies the power levels within the sensor node depending on the non-deterministic workload. By varying the voltage along with the frequency, it is possible to obtain quadratic reduction in power consumption.

Sensors

Sensors are used by wireless sensor nodes to capture data from their environment. They are hardware devices that produce a measurable response to a change in a physical condition like temperature or pressure. Sensors measure physical data of the parameter to be monitored and have specific characteristics such as accuracy, sensitivity etc. The continual analog signal produced by the sensors is digitized by an analog-to-digital converter and sent to controllers for further processing. Some sensors contain the necessary electronics to convert the raw signals into readings which can be retrieved via a digital link (e.g. I2C, SPI) and many convert to units such as °C. Most sensor nodes are small in size, consume little energy, operate in high volumetric densities, be autonomous and operate unattended, and be adaptive to the environment. As wireless sensor nodes are typically very small electronic devices, they can only be equipped with a limited power source of less than 0.5-2 ampere-hour and 1.2-3.7 volts.

Sensors are classified into three categories: passive, omnidirectional sensors; passive, narrow-beam sensors; and active sensors. Passive sensors sense the data without actually manipulating the environment by active probing. They are self powered; that is, energy is needed only to amplify their analog signal. Active sensors actively probe the environment, for example, a sonar or radar sensor, and they require continuous energy from a power source. Narrow-beam sensors have a well-defined notion of direction of measurement, similar to a camera. Omnidirectional sensors have no notion of direction involved in their measurements.

Most theoretical work on WSNs assumes the use of passive, omnidirectional sensors. Each sensor node has a certain area of coverage for which it can reliably and accurately report the particular quantity that it is observing. Several sources of power consumption in sensors are: signal sampling and conversion of physical signals to electrical ones, signal conditioning, and analog-to-digital conversion. Spatial density of sensor nodes in the field may be as high as 20 nodes per cubic meter.

Mobile Wireless Sensor Network

A mobile wireless sensor network (MWSN) can simply be defined as a wireless sensor network (WSN) in which the sensor nodes are mobile. MWSNs are a smaller, emerging field of research in contrast to their well-established predecessor. MWSNs are much more versatile than static sensor networks as they can be deployed in any scenario and cope with rapid topology changes. However, many of their applications are similar, such as environment monitoring or surveillance. Commonly, the nodes consist of a radio transceiver and a microcontroller powered by a battery, as well as some kind of sensor for detecting light, heat, humidity, temperature, etc.

Challenges

Broadly speaking, there are two sets of challenges in MWSNs; hardware and environment. The main hardware constraints are limited battery power and low cost requirements. The limited power means that it's important for the nodes to be energy efficient. Price limitations often demand low complexity algorithms for simpler microcontrollers and use of only a simplex radio. The major environmental factors are the shared medium and varying topology. The shared medium dictates that channel access must be regulated in some way. This is often done using a medium access control (MAC) scheme, such as carrier sense multiple access (CSMA), frequency division multiple access (FDMA) or code division multiple access (CDMA). The varying topology of the network comes from the mobility of nodes, which means that multihop paths from the sensors to the sink are not stable.

Standards

Currently there is no standard for MWSNs, so often protocols from MANETs are borrowed, such as Associativity-Based Routing (AR), Ad hoc On-Demand Distance Vector Routing (AODV), Dynamic Source Routing (DSR) and Greedy Perimeter Stateless Routing (GPSR). MANET protocols are preferred as they are able to work in mobile environments, whereas WSN protocols often aren't suitable.

Topology

Topology selection plays an important role in routing because the network topology decides the transmission path of the data packets to reach the proper destination. Here, all the topologies (Flat / Unstructured, cluster, tree, chain and hybrid topology) are not feasible for reliable data transmission on sensor nodes mobility. Instead of single topology, hybrid topology plays a vital role in data collection, and the performance is good. Hybrid topology management schemes include the Cluster Independent Data Collection Tree (CIDT). and the Velocity Energy-efficient and Link-aware Cluster-Tree (VELCT); both have been proposed for mobile wireless sensor networks (MWSNs).

Routing

Since there is no fixed topology in these networks, one of the greatest challenges is routing data from its source to the destination. Generally these routing protocols draw inspiration from two fields; WSNs and mobile ad hoc networks (MANETs). WSN routing protocols provide the required functionality but cannot handle the high frequency of topology changes. Whereas, MANET routing protocols can deal with mobility in the network but they are designed for two way communication, which in sensor networks is often not required.

Protocols designed specifically for MWSNs are almost always multihop and sometimes adaptations of existing protocols. For example, Angle-based Dynamic Source Routing (ADSR), is an adaptation of the wireless mesh network protocol Dynamic Source Routing (DSR) for MWSNs. ADSR uses location information to work out the angle between the node intending to transmit, potential forwarding nodes and the sink. This is then used to insure that packets are always forwarded towards the sink. Also, Low Energy Adaptive Clustering Hierarchy (LEACH) protocol for WSNs has been adapted to LEACH-M (LEACH-Mobile), for MWSNs. The main issue with hierarchical protocols is that mobile nodes are prone to frequently switching between clusters, which can cause large amounts of overhead from the nodes having to regularly re-associate themselves with different cluster heads.

Another popular routing technique is to utilise location information from a GPS module attached to the nodes. This can be seen in protocols such as Zone Based Routing (ZBR), which defines clusters geographically and uses the location information to keep nodes updated with the cluster they're in. In comparison, Geographically Opportunistic Routing (GOR), is a flat protocol that divides the network area into grids and then uses the location information to opportunistically forward data as far as possible in each hop.

Multipath protocols provide a robust mechanism for routing and therefore seem like a promising direction for MWSN routing protocols. One such protocol is the query based Data Centric Braided Multipath (DCBM).

Medium Access Control

There are three types of medium access control (MAC) techniques: based on time division, frequency division and code division. Due to the relative ease of implementation, the most common choice of MAC is time-division-based, closely related to the popular CSMA/CA MAC.

Validation

Protocols designed for MWSNs are usually validated with the use of either analytical, simulation or experimental results. Detailed analytical results are mathematical in nature and can provide good approximations of protocol behaviour. Simulations can be performed using software such as OPNET, NetSim and NS2 and is the most common method of validation. Simulations can provide close approximations to the real behaviour of a protocol under various scenarios. Physical experiments are the most expensive to perform and, unlike the other two methods, no assumptions need to be made. This makes them the most reliable form of information, when determining how a protocol will perform under certain conditions.

Applications

The advantage of allowing the sensors to be mobile increases the number of applications beyond those for which static WSNs are used. Sensors can be attached to a number of platforms:

- People

- Animals

- Autonomous Vehicles

- Unmanned Vehicles

- Manned Vehicles

In order to characterise the requirements of an application, it can be categorised as either constant monitoring, event monitoring, constant mapping or event mapping. Constant type applications are time-based and as such data is generated periodically, whereas event type applications are event drive and so data is only generated when an event occurs. The monitoring applications are constantly running over a period of time, whereas mapping applications are usually deployed once in order to assess the current state of a phenomenon. Examples of applications include health monitoring, which may include heart rate, blood pressure etc. This can be constant, in the case of a patient in a hospital, or event driven in the case of a wearable sensor that automatically reports your location to an ambulance team in the case of an emergency. Animals can have sensors attached to them in order to track their movements for migration patterns, feeding habits or other research purposes. Sensors may also be attached to unmanned aerial vehicles (UAVs) for surveillance or environment mapping. In the case of autonomous UAV aided search and rescue, this would be considered an event mapping application, since the UAVs are deployed to search an area but will only transmit data back when a person has been found.

Optical Wireless Communications

Optical wireless communications (OWC) is a form of optical communication in which unguided visible, infrared (IR), or ultraviolet (UV) light is used to carry a signal.

OWC systems operating in the visible band (390–750 nm) are commonly referred to as visible light communication (VLC). VLC systems take advantage of light emitting diodes (LEDs) which can be pulsed at very high speeds without noticeable effect on the lighting output and human eye. VLC can be possibly used in a wide range of applications including wireless local area networks, wireless personal area networks and vehicular networks among others. On the other hand, terrestrial point-to-point OWC systems, also known as the free space optical (FSO) systems, operate at the near IR frequencies (750–1600 nm). These systems typically use laser transmitters and offer a cost-effective protocol-transparent link with high data rates, i.e., 10 Gbit/s per wavelength, and provide a potential solution for the backhaul bottleneck. There has also been a growing interest on ultraviolet communication (UVC) as a result of recent progress in solid state optical sources/detectors operating within solar-blind UV spectrum (200–280 nm). In this so-called deep UV band, solar radiation is negligible at the ground level and this makes possible the design of photon-counting detectors with wide field-of-view receivers that increase the received energy with little additional background noise. Such designs are particularly useful for outdoor non-line-of-sight configurations to support low power short-range UVC such as in wireless sensor and ad-hoc networks.

History

The proliferation of wireless communications stands out as one of the most significant phenomena in the history of technology. Wireless technologies have become essential much more quickly during the last four decades and they will be a key element of society progress for the foreseeable future. The radio-frequency (RF) technologies wide-scale deployment has become the key factor to the wireless devices and systems expansion. However, the electromagnetic spectrum where the wireless systems are deployed is limited in capacity and costly according to its exclusive licenses of exploitation. With the raise of data heavy wireless communications, the demand for RF spectrum is outstripping supply and they become to consider other viable options for wireless communication using the upper parts of the electromagnetic spectrum not just RF.

Optical wireless communication (OWC) refers to transmission in unguided propagation media through the use of optical carriers, i.e., visible, infrared (IR), and ultraviolet (UV) band. Signalling through beacon fires, smoke, ship flags and semaphore telegraph can be considered the historical forms of OWC. Sunlight has been also used for long distance signalling since very early times. The earliest use of sunlight for communication purposes is attributed to ancient Greeks and Romans who used their polished shields to send signals by reflecting sunlight during battles. In 1810, Carl Friedrich Gauss invented the heliograph which involves a pair of mirrors to direct a controlled beam of sunlight to a distant station. Although the original heliograph was designed for geodetic survey, it was used extensively for military purposes during the late 19th and early 20th century. In 1880, Alexander Graham Bell invented the photophone, known as the world's first wireless telephone system.

The military interest on photophone however continued. For example, in 1935, the German Army developed a photophone where a tungsten filament lamp with an IR transmitting filter was used as a light source. Also, American and German military laboratories continued the development of high pressure arc lamps for optical communication until the 1950s. In modern sense, OWC uses either lasers or light emitting diodes (LEDs) as transmitters. In 1962, MIT Lincoln Labs built an experimental OWC link using a light emitting GaAs diode and was able to transmit TV signals over a distance of 30 miles. After the invention of laser, OWC was envisioned to be the main deployment area for lasers and many trials were conducted using different types of lasers and modulation schemes. However, the results were in general disappointing due to large divergence of laser beams and the inability to cope with atmospheric effects. With the development of low-loss fiber optics in the 1970s, they became the obvious choice for long distance optical transmission and shifted the focus away from OWC systems.

Current Status

Over the decades, the interest in OWC remained mainly limited to covert military applications, and space applications including inter-satellite and deep-space links.

OWC's mass market penetration has been so far limited with the exception of IrDA which became a highly successful wireless short-range transmission solution. Development of novel and efficient wireless technologies for a range of transmission links is essential for building future heterogeneous communication networks to support a wide range of service types with various traffic patterns and to meet the ever-increasing demands for higher data rates. Variations of OWC can be potentially employed in a diverse range of communication applications ranging from optical interconnects within integrated circuits through outdoor inter-building links to satellite communications.

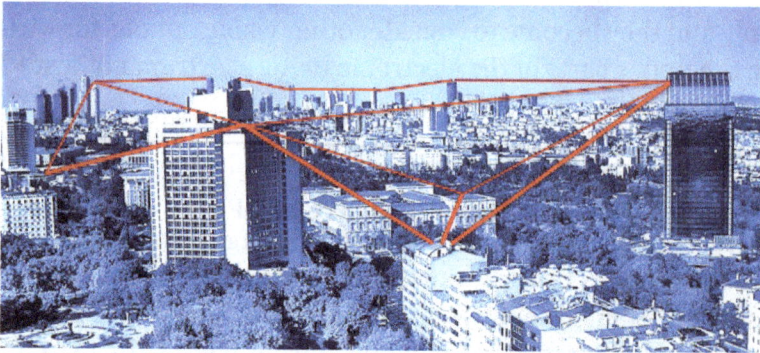

OWC long range inter-buildings communications idea for Istanbul Skyline

Applications

Based on the transmission range, OWC can be studied in five categories:

1. Ultra-short range OWC: chip-to-chip communications in stacked and closely packed multi-chip packages.

2. Short range OWC: wireless body area network (WBAN) and wireless personal area network (WPAN) applications under standard IEEE 802.15.7, underwater communications.

3. Medium range OWC: indoor IR and visible light communications (VLC) for wireless local area networks (WLANs) and inter-vehicular and vehicle-to-infrastructure communications.

4. Long range OWC,: inter-building connections, also called Free-Space Optical Communications (FSO).

5. Ultra-long range OWC: inter-satellite links.

Recent Trends

- In January 2015, IEEE 802.15 formed a Task Group to write a revision to IEEE 802.15.7-2011 that accommodates infrared and near ultraviolet wavelengths, in addition to visible light, and adds options such as Optical Camera Communications and LiFi.

- At long range OWC applications a 1 Gbit/s - 60 km range link between ground to aircraft at 800 km/h speed has been demonstrated, "Extreme Test for the ViaLight Laser Communication Terminal MLT-20 – Optical Downlink from a Jet Aircraft at 800 km/h", DLR and EADS December 2013.

- On consumer devices and short-range OWC applications on phones; Charge and receive data with light at your smartphone: TCL Communication/ALCATEL ONETOUCH and Sunpartner Technologies announces the first fully integrated solar smartphone. March 2014.

- On ultra-long range OWC applications the NASA's Lunar Laser Communication Demonstration (LLCD) transmitted data from lunar orbit to Earth at a rate of 622 Megabits-per-second (Mbps), November 2013.

- The Next Generation of OWC / Visible Light Communications demonstrated 10 Mb/s transmission with Polymer Light-Emitting Diodes or OLED.

- On OWC research activities there is a European research project action IC1101 OPTICWISE of the COST Programme (European Cooperation in Science and Technology) funded by the European Science Foundation, allowing the coordination of nationally funded research on a European level. The Action aims to serve as a high-profile consolidated European scientific platform for interdisciplinary optical wireless communication (OWC) research activities. It was launched in November 2011 and will run until November 2015. More than 20 countries represented.

- The consumer and industry OWC technologies adoption is represented by the Li-Fi Consortium, founded in 2011 is a Non-profit organization, devoted to introduce optical wireless technology. Promotes the adoption of Light Fidelity (Li-Fi) products.

- An example of Asian awareness about OWC is the VLCC visible light communication consortium in Japan, established at 2007 in order to realize safe, ubiquitous telecommunication system using visible light through the activities of market research, promotion, and standardization.

- In the USA there are several OWC initiatives, including the "Smart Lighting Engineering Research Center", founded in 2008 by the National Science Foundation (NSF) is a partnership of Rensselaer Polytechnic Institute (lead institution), Boston University and the University of New Mexico. Outreach partners are Howard University, Morgan State University, and Rose-Hulman Institute of Technology.

References

- Dargie, W. and Poellabauer, C. (2010). Fundamentals of wireless sensor networks: theory and practice. John Wiley and Sons. pp. 168–183, 191–192. ISBN 978-0-470-99765-9

- Magno, M.; Boyle, D.; Brunelli, D.; O'Flynn, B.; Popovici, E.; Benini, L. (2014). "Extended Wireless Monitoring Through Intelligent Hybrid Energy Supply". IEEE Transactions on Industrial Electronics. 61 (4): 1871. doi:10.1109/TIE.2013.2267694

- K. Saleem; N. Fisal & J. Al-Muhtadi (2014). "Empirical studies of bio-inspired self-organized secure autonomousRouting protocol". Sensors Journal IEEE. 14: 1–8. doi:10.1109/JSEN.2014.2308725

- Sohraby, K., Minoli, D., Znati, T. (2007). Wireless sensor networks: technology, protocols, and applications. John Wiley and Sons. pp. 203–209. ISBN 978-0-471-74300-2

- Spie (2013). "Vassili Karanassios: Energy scavenging to power remote sensors". SPIE Newsroom. doi:10.1117/2.3201305.05

- B. White et al. 2008. Contaminant Cloud Boundary Monitoring Using Network of UAV Sensors. IEEE Sensors Journal, vol. 8, no. 10, pp. 1681-1692

- Guowang Miao; Jens Zander; Ki Won Sung; Ben Slimane (2016). Fundamentals of Mobile Data Networks. Cambridge University Press. ISBN 1107143217

- Survey on Centralised and Distributed Clustering Routing Algorithms for WSNs (PDF). IEEE 81st Vehicular Technology Conference. Glasgow, Scotland: IEEE. Spring 2015. doi:10.1109/VTC-Spring.2015.7145650. Retrieved March 4, 2016

- R. Velmani, and B. Kaarthick, 2015. An Efficient Cluster-Tree Based Data Collection Scheme for Large Mobile Wireless Sensor Networks. IEEE Sensors Journal, vol. 15, no. 4, pp. 2377–2390. doi: 10.1109/JSEN.2014.2377200

- T. Hayes and F.H. Ali. 2016. "Mobile Wireless Sensor Networks: Applications and Routing Protocols". Handbook of Research on Next Generation Mobile Communications Systems. IGI Global. ISBN 9781466687325. pp.256-292

- I. F. Akyildiz and I.H. Kasimoglu (2004). "Wireless Sensor and Actor Networks: Research Challenges". Ad Hoc Networks. 2 (4): 351–367. doi:10.1016/j.adhoc.2004.04.003

- Muaz Niazi, Amir Hussain (2011). A Novel Agent-Based Simulation Framework for Sensing in Complex Adaptive Environments. IEEE Sensors Journal, Vol.11 No. 2, 404–412. Paper

- Aghdam, Shahin Mahdizadeh; Khansari, Mohammad; Rabiee, Hamid R; Salehi, Mostafa (2014). "WCCP: A congestion control protocol for wireless multimedia communication in sensor networks". Ad Hoc Networks. 13: 516–534. doi:10.1016/j.adhoc.2013.10.006

5

Various Wireless Devices and Hardware

Various devices that enable wireless networking include wireless router, MiFi, Mobile broadband modem, etc. A wireless router sends data packets using WLAN as well as provides access to a wired network. Tools and techniques are an important component of any field of study. The following chapter elucidates the various tools and techniques that are related to WiFi.

Wireless Router

A wireless router is a device that performs the functions of a router and also includes the functions of a wireless access point. It is used to provide access to the Internet or a private computer network. It can function in a wired LAN (local area network), in a wireless-only LAN (WLAN), or in a mixed wired/wireless network, depending on the manufacturer and model.

Features

Most current wireless routers have the following characteristics:

- One or multiple NICs supporting Fast Ethernet or Gigabit Ethernet integrated into the main SoC

- One or multiple WNICs supporting a part of the IEEE 802.11-standard family also integrated into the main SoC or as separate chips on the printed circuit board. It also can be a distinct card connected over a MiniPCI or MiniPCIe interface.

  - So far the PHY-Chips for the WNICs are generally distinct chips on the PCB. Dependent on the mode the WNIC supports, i.e. 1T1R, 2T2R or 3T3R, one WNIC have up to 3 PHY-Chips connected to it. Each PHY-Chip is connected to a Hirose U.FL-connector on the PCB. A so-called pigtail cable connects the Hirose U.FL either to a RF connector, in which case the antenna can be changed or directly to the antenna, in which case it is integrated into the casing. Common are single-band (i.e. only for 2.4 GHz or only for 5 GHz) and dual-band (i.e. for 2.4 and 5 GHz) antennas.

- Often an Ethernet switch supporting Gigabit Ethernet or Fast Ethernet, with support for IEEE 802.1Q, integrated into the main SoC (MediaTek SoCs) or as separate Chip on the PCB.

- Some wireless routers come with either xDSL modem, DOCSIS modem, LTE modem, or fiber optic modem integrated.

- IEEE 802.11n compliant or ready.

- Some dual-band wireless routers operate the 2.4 GHz and 5 GHz bands simultaneously.

- Some high end dual-band wireless routers have data transfer rates of at most 300 Mbit/s (For 2.4 GHz band) and 450 Mbit/s (For 5 GHz band).

- The Wi-Fi clone button simplifies Wi-Fi configuration and builds a seamless unified home network, enabling Super Range Extension, which means it can automatically copy the SSID and Password of your router.

- Some wireless routers have one or two USB ports. For wireless routers having one USB port, it is designated for either printer or desktop/mobile external hard disk drive. For wireless routers having two USB ports, one is designated for the printer and the other one is designated for either desktop or mobile external hard disk drive.

- Some wireless routers have a USB port specifically designed for connecting mobile broadband modem, aside from connecting the wireless router to an Ethernet with xDSL or cable modem. So, can be inserted a mobile broadband USB adapter into the router to share the mobile broadband Internet connection through the wireless network.

Notable Manufacturers

• Apple Inc.	• Linksys
• Arris	• MikroTik
• Belkin	• Motorola
• Buffalo Technology	• Netgear
• Cisco	• TP-Link
• D-Link	
• HP Inc.	• Ubiquiti Networks

Operating System

The most common operating system on such embedded devices is Linux. More seldomly, VxWorks is being used. The devices are configured over a web user interface served by a light web server software running on the device.

It is possible for a computer running a desktop operating system such as Windows to, with appropriate software, act as a wireless router. This is commonly referred to as a SoftAP, or "Software Access Point".

Open Source Firmware

In 2003, Linksys was forced to open-source the firmware of its WRT54G router after people on the Linux Kernel Mailing List discovered that it used GPL Linux code. In 2008, Cisco was sued in *Free Software Foundation, Inc. v. Cisco Systems, Inc* due to similar issues with Linksys routers.

Since then, various open-source projects have built on this foundation, including Open-Wrt, DD-WRT, and Tomato.

In 2016, various manufacturers changed their firmware to block custom installations after an FCC ruling. However, some companies plan to continue to officially support open-source firmware, including Linksys and Asus.

MiFi

A Novatel MiFi 2372 "Intelligent Mobile Wi-Fi Hotspot"

MiFi is a brand name used to describe a wireless router that acts as mobile Wi-Fi hotspot. In many countries, including the United States, Canada, and Mexico, Novatel Wireless (now known as Inseego Corp.) owns a registered trademark on the "MiFi" brand name; in the United Kingdom mobile operator Hutchison 3G owns the "MiFi" trademark. Novatel Wireless has never offered an official explanation for the origin of the name "MiFi"; it is believed to be short for "My Wi-Fi". In September 2016 Novatel Wireless announced that it has agreed to sell the MiFi brand to TCL Industries Holdings of Hong Kong; the sale is expected to close in early 2017, pending approval from shareholders and regulators.

A MiFi device can be connected to a cellular network and provide internet access for up to ten devices. Novatel Wireless introduced the first MiFi device in the United States, in May 2009. 3' "MiFi" is a similar line from Huawei under the name.

MiFi Brand Name

Novatel Wireless owns a registered trademark on the "MiFi" brand name in the U.S. (including Puerto Rico), and a number of countries worldwide: Bahrain, Canada, Egypt, Germany, Ghana, Hungary, Japan, Kuwait, Mexico, Pakistan, the Netherlands, New Zealand, Poland, Portugal, Qatar, Romania, Singapore, Slovenia, South Africa, Spain, and Thailand.

The notable exception is in the UK where mobile operator 3 owns the "MiFi" trademark. In India the Mi-Fi trademark is owned by Mi-Fi Networks Private Limited

Devices

Novatel Wireless MiFi 2200

MiFi 2200 from Novatel Wireless for Verizon Wireless

- Limited to five Wi-Fi clients such as laptops, cameras, gaming devices, and multimedia players; with the exception of manually editing the device's config file to allow more clients.

- May be connected to a computer via a Micro-USB connection, though doing so disables the Wi-Fi networking, converting the device into a traditional single-client modem. (However, CNET has introduced a tweak to charge the device over USB while maintaining its functionality.)

- Includes GPS unit, which is usable on some networks like Virgin Mobile and not on others like Verizon.

- Uses 3G data network (CDMA 1xEVDO RevA).

Novatel Wireless MiFi 23xx series

Same functionality as 2200, plus:

- Accepts SD card for in-device shared media storage.

- Uses 3G data network (MiFi 2352: HSUPA/HSDPA 900/1900/2100 MHz, MiFi 2372: HSUPA/HSDPA 850/1900/2100 MHz; both support GPRS/EDGE 850/900/1800/1900 MHz).

Novatel Wireless MiFi 33xx series

Same functionality as 23xx series, plus:

- Linux-based MiFi OS with widgets:

  o Messages: perform SMS-based messaging actions such as reading, writing, sending, and receiving SMS messages

  o Data usage: track MiFi data usage in home and roaming networks

  o GeoSearch: leverage the GPS functionality of the MiFi to display a map of the local area, search the local area, display the search results on the map

  o Weather: fetch weather data for the current and defined locations

  o MiFi DLNA Server: start, stop, and configure the MiFi DLNA server

4G Mobile Hotspot Devices

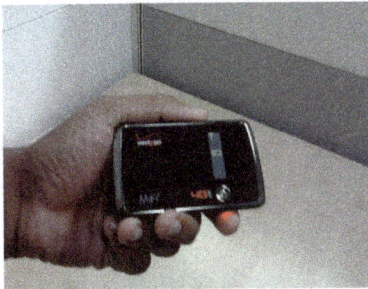

MiFi 4510L from Novatel Wireless for
Verizon Wireless

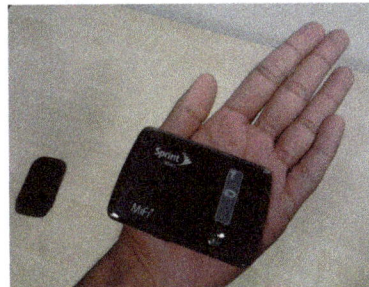

MiFi 4082 from Novatel Wireless for
Sprint Nextel

The Las Vegas Consumer Electronics Show 2011 saw the introduction of two new 4G-capable MiFi devices from Novatel:

- Verizon Wireless featured the 4510L model, which will connect to Verizon's LTE 4G network, expected to support 5–12 Mbps download and 2–5 Mbit/s upload.

- Sprint featured the MiFi 4082, a WiMAX version.

Both devices maintain backward compatibility with existing 3G networks. Other features include:

- MiFiOS with widget support

- MicroSD card slot

- GPS receiver

- Four-hour battery life

- Front panel status display using E Ink technology. The front panel display shows battery, signal strength, and number of connected devices. The difficulty of viewing such information was seen as a major shortcoming of earlier MiFi devices.

Non-Novatel Wireless Devices

A number of providers other than Novatel provide personal hotspot, "MiFi"-like services:

nyx mobile "mifi LTE" mobile router

- Alcatel One Touch Link Y800 sold through EE in the UK

- D-Link DIR-457/MyPocket

- Freedom Spot Personal Hotspot

- Goodspeed mobile hotspot supports 3G/3.5G

- Huawei E5 E5830 (Series), E585, E586 with HSPA+ and Chinese market E5805 using CDMA2000 and ET536 using TD-SCDMA

- mifi LTE nyx mobile is a mobile router 3G + 4G (LTE) launched by nyx mobile for Telcel. It is built using Qualcomm technology and is expected to support up to 100 Mbps download. (Announced by June 2013 for México

- Netgear AirCard 781S (Sold under the name Zing by Sprint)

- Option GlobeSurfer III

- Packet One Networks MF230 (offered as part of their P1 ToGo plan]

- Sierra Wireless Overdrive (4G capable; available only in the United States through Sprint)

- ZTE MF60, MF80

- JioFi is a mobile router 4G (LTE) launched in India.

- Rentafi is Singapore's leading portable 4G WiFi rental service.

Alternatives

Mobile phones with an Internet connection can often be turned into Wi-Fi hotspots using a process called "tethering", which is similar to using dedicated MiFi devices.

The following phone families have built-in features to create Wi-Fi access point:

- Android phones running Android 2.2 or later

- BlackBerry devices running OS 7.1 or later

- iPhone 3GS and iPhone 4 running iOS 4.3 or later, and all iPhones on Verizon Wireless (initially released with iOS 4.2.5)

- Palm Pixi Plus and Pre Plus on Verizon Wireless, with a 5GB cap

- Windows Phone devices running OS 7.5 or later (and if allowed by operator)

For other phones there are third-party applications to allow this:

- Android running Android 2.1 or under – Wireless Tether

- iPhone 3G and later – MyWi (requires jailbreak).

- PiFi – Personal WiFi Device

- S60 phones and Nokia N900 – JoikuSpot

- Windows Mobile – WMWifiRouter

Awards

- Novatel MiFi 2200:

  o Mobile Village "Mobile Star" (Portable Remote Connectivity Gear): "Super-star" Award (2009)

  o Laptop Magazine Editor's Choice (2009)

- PC World "Gear of the Year" (2009)

- Mobile News "Most Innovative Product (non-handset)" (2010)

- Novatel MiFi 2352:

  - Plus X Award in Technology (2009)

  - CTIA Emerging Technology Award (Fashion & Lifestyle Products), 1st Place (2009)

  - CTIA "Hot for the Holidays" (Mobile Internet Device or Netbook) Award (2009)

  - CES Innovations Award (Enabling Technologies) Winner (2010)

  - Mobile World Congress "Global Mobile" Award: Best Mobile Connected Device (2010)

- Novatel 4G MiFi

  - Consumer Electronics Show 2011, Notebook Accessories Category: Best in Show

  - 2010 World Communication Award (WCA) for Best Mobile Device Strategy

  - Mobile Village Mobile Star Award for Best Laptop or Tablet Accessory

MiFi in the News

Security Issues

In January 2010, two major security holes were discovered with the Novatel MiFi 2200 which, if properly exploited, could allow a malicious user to obtain the device's current GPS location and security keys. If the malicious user were physically close enough to use the device's Wi-Fi signal, this could give access to the MiFi's 3G connection as well as any other connected devices. Novatel responded that a security patch would be available in February 2010.

The popularity of MiFi devices can also be problematic for corporate network security. Corporations generally expect to control on-site internet access: many use firewalls to reduce the risk of malware, and some enforce restrictions aimed at employee productivity. Personal mobile hotspots may provide a "back door" by which employees can circumvent these precautions.

Recall

In May 2010, the Mifi 2372 was recalled in Canada by Bell Mobility and Rogers Communications. In two documented cases, difficulty of opening the MiFi battery compartment had caused customers to use levels of force that caused physical damage to the batteries, which subsequently overheated. Novatel replaced the recalled units with a type that featured an easier-to-open battery compartment.

Customers were sent prepaid courier envelopes and instructed to send back their Mifi units for replacement of the battery and battery compartment cover. Customers were notified that Novatel would return the serviced units within 6 to 8 weeks of their return. Bell customers were provided with cellular internet access via Novatel U998 USB sticks, which were provided as temporary replacements by Bell.

Radio Interference at Trade Shows

At two major trade shows in 2010— Google's first public demo of Google TV and the iPhone 4 demonstrations at the 2010 Apple Worldwide Developers Conference— keynote presentations using available Wi-Fi connectivity were disrupted by network unreliability. The problem was traced to massive radio interference, caused by the popularity of MiFi and similar devices for "liveblogging" from the trade show floor. In the case of the Apple conference, Apple CEO Steve Jobs stated that 570 different Wi-Fi networks ("several hundred" being MiFis) were operating simultaneously in the exhibit hall.

Mobile Broadband Modem

The Huawei E1691 sold at Wind Mobile in Canada is a USB modem providing a mobile broadband connection.

A mobile broadband modem, also known as a *connect card* or *data card*, is a type of modem that allows a laptop, a personal computer or a router to receive Internet access via a mobile broadband connection instead of using telephone or cable television lines. A mobile Internet user can connect using a wireless modem to a wireless Internet Service Provider (ISP) to get Internet access.

History

1G and 2G

While some analogue mobile phones provided a standard RJ11 telephone socket into which a normal landline modem could be plugged, this only provided slow dial-up connections, usually 2.4 kilobit per second (kbit/s) or less. The next generation of phones, known as 2G (for 'second generation'), were digital, and offered faster dial-up speeds of 9.6kbit/s or 14.4kbit/s without the need for a separate modem. A further evolution called HSCSD used multiple GSM channels (two or three in each direction) to support up to 43.2kbit/s. All of these technologies still required their users to have a dial-up ISP to connect to and provide the Internet access - it was not provided by the mobile phone network itself.

The release of 2.5G phones with support for packet data changed this. The 2.5G networks break both digital voice and data into small chunks, and mix both onto the network simultaneously in a process called packet switching. This allows the phone to have a voice connection and a data connection at the same time, rather than a single channel that has to be used for one or the other. The network can link the data connection into a company network, but for most users the connection is to the Internet. This allows web browsing on the phone, but a PC can also tap into this service if it connects to the phone. The PC needs to send a special telephone number to the phone to get access to the packet data connection. From the PC's viewpoint, the connection still looks like a normal PPP dial-up link, but it is all terminating on the phone, which then handles the exchange of data with the network. Speeds on 2.5G networks are usually in the 30–50kbit/s range.

3G

Huawei CDMA2000 Evolution-Data Optimized USB wireless modem

3G networks have taken this approach to a higher level, using different underlying technology but the same principles. They routinely provide speeds over 300kbit/s. Due to the now increased internet speed, internet connection sharing via WLAN has become a workable reality. Devices which allow internet connection sharing or other types of routing on cellular networks are called also cellular routers.

A further evolution is the 3.5G technology HSDPA, which provides speeds of multiple Megabits per second. Several of the mobile network operators that provide 3G or faster wireless internet access offer plans and wireless modems that enable computers to connect to and access the internet. These wireless modems are typically in the form of a small USB based device or a small, portable mobile hotspot that acts as a WiFi access point (hotspot) to enable multiple devices to connect to the internet. WiMAX based services that provide high speed wireless internet access are available in some countries and also rely on wireless modems that connect to the provider's wireless network. Wireless USB Modems are nicknamed as "Dongles".

Early 3G mobile broadband modems used the PCMCIA or ExpressCard ports, commonly found on legacy laptops. The expression "connect card" (instead of connection card) had been registered and used the first time by Vodafone as brand for its products but now is become a brandnomer or genericized trademark used in colloquial or commercial speech for similar product, made by different manufacturers, too. Major producers are Huawei, Option N.V., Novatel Wireless. More recently, the expression "connect card" is also used to identify internet USB keys. Vodafone brands this type of device as a Vodem.

Often a mobile network operator will supply a 'locked' modem or other wireless device that can only be used on their network. It is possible to use online unlocking services that will remove the 'lock' so the device accepts SIM cards from any network.

Variants

HSDPA cellular router

Standalone

Standalone mobile broadband modems are designed to be connected directly to one computer. In the past the PCMCIA and ExpressCard standards were used to connect to the computer. As USB connectivity became almost universal, these various standards were largely superseded by USB modems in the early 21st century. Some models have GPS support, providing geographical location information.

Integrated Router

Many mobile broadband modems sold nowadays also have built-in routing capabilities. They provide traditional networking interfaces such as Ethernet, USB and Wi-Fi.
Models are available for both consumers and enterprises. Some require the use of an
AC adapter, while others are portable and can also be powered by a USB connection
or a built-in battery. An RJ11 registered jack is also present on a few of these modems,
allowing the connection of a traditional home phone to make cellular calls.

Smartphones and Tethering

Numerous smartphones support the Hayes command set and therefore can be used as a
mobile broadband modem. Some mobile network operators charge a fee for this facility,
if able to detect the tethering. Other networks have an allowance for full speed mobile
broadband access, which—if exceeded—can result in overage charges or slower speeds.

An Internet-accessing smartphone may have the same capabilities as a standalone modem, and, when connected via a USB cable to a computer, can serve as a modem for the
computer. Smartphones with built-in Wi-Fi also typically provide routing and wireless
access point facilities. This method of connecting is commonly referred to as "tethering."

Service Providers

There are competing common carriers broadcasting signal in most nations of the earth.

Technologies

• HSPA+ 3.75G	• CDPD
• DC-HSPA+	• CDMA2000 (3G)
• iBurst (pre-4G)	• EDGE
• HiperMAN (pre-4G)	• EVDO (3G, although could be considered to be 3.5G due to its peak bandwidth.)
• WiMAX (pre-4G)	
• WiBro (pre-4G)	• UMTS (3G)
• LTE (4G)	• GPRS Core Network
• LTE Advanced (4G)	• IP Multimedia Subsystem
• GPRS (2.5G)	• HSDPA (3.5G)

Device Driver Switching

Mobile broadband modems often use a virtual CD-ROM switching utility and have the
device drivers on board. Those modems have two modes, a USB flash drive mode and
in the other mode they are a modem.Via the USB Protocol.

WokFi

Mid-sized WokFi antenna sample

WokFi (a portmanteau derived from blending the words Wok + Wi-Fi) is a slang term for a style of homemade Wi-Fi antenna consisting of a crude parabolic antenna made with a low-cost Asian kitchen wok, spider skimmer or similar household metallic dish. The dish forms a directional antenna which is pointed at the wireless access point antenna, allowing reception of the wireless signal at greater distances than standard omnidirectional Wi-Fi antennas.

Description

WokFi antennas are fabricated out of commonly available concave metal kitchen dishes or dish covers (which need not be perfectly parabolic); Asian woks are favored because they have shapes closest to parabolic. A commercial Wi-Fi antenna, usually a USB Wi-Fi dongle, is suspended in front of the dish, attached by cable to the computer.

The WokFi antenna is considered simpler and cheaper than other home-built antenna projects (such as the popular cantenna), but is a very effective method to boost the Wi-Fi connection quality, audit access point coverage, and even quickly establish WLAN viability – perhaps if a more professional setup is eventually intended.

Advantages

A significant advantage is that with a USB modem the RF signal is converted to a conventional digital signal at the antenna. Therefore, by using standard USB extension cables, the antenna can be located at a distance from the computer of five meters or more, with no concerns over microwave signal losses that would occur in an RF coaxial cable feedline of that length used to attach a conventional antenna to the RF input of a computer modem. Chaining active USB repeaters, it is possible to locate the antenna at much greater distances from the computer, which is especially useful when line-of-sight (LOS) obstacles (such as vegetation and walls) require the antenna to be located on the roof, for example. If using mesh reflectors, usually with a grid under 5 mm, the antenna will be lighter and less wind-prone than larger dishes.

Performance

WokFi gains are typically 10+ dB, with range boosts, thus can be 16-32 times over a bare USB adapter. Ranges (LoS) are typically 3–5 km, although an aligned pair of similar point-to-point transceiver setups may approach 10 km over a clear path. In addition, certain improved WokFi antennas, and antennas made using surplus 60 to 90 cm (2-3 ft) diameter round or oval satellite TV dishes, allow even far greater range, up to 20 km.

Interference from nearby 2.4 GHz signals (perhaps from cordless phones, AV links, leaky microwave ovens, other APs or Bluetooth) can be nulled out—a useful feature in this increasingly crowded part of the RF spectrum. The performance of abundant, low-powered Wi-Fi "dongles", typically selling for approximately US$15–20, but of only 30–40 mW transmitter power and modest receiver sensitivity, can easily be boosted with little more than cheap cookware or pot lids. The "sweet spot" on such ad hoc reflectors can readily be found by taping a small (~2.5 cm, or 1 in) mirror on the surface of the dish, to see where the sun's rays focus.

Some Other Examples

WiFi Master Key

WiFi Master Key is a product of the Shanghai LinkSure Limited Company founded in 2013; LinkSure is a mobile internet company providing free internet and content services. LinkSure's core product is WiFi Master Key, which leverages the sharing economy, cloud computing, and big data in an attempt to provide users with a safe and free Wi-Fi internet connection shared by Wi-Fi hosts around the world. The company's Chairman and CEO, Mr. Chen Danian, was previously COO and co-founder of Shanda Games.

At the beginning of 2015, LinkSure closed its A round funding of USD 52 million at a billion-dollar valuation, becoming a mobile internet unicorn.

In May 2015, LinkSure bought the domain name wifi.com and established a branch in Singapore to expand its overseas services. As of June 2016, WiFi Master Key's overseas users have exceeded 80 million, with over 20 million daily active users, covering 223 countries and regions across 5 continents. In 49 countries, including Russia, Brazil, Indonesia, Thailand, Vietnam, Hong Kong, and Taiwan, they are currently ranked number 1 on the tools section of the Google Play store. According to App Annie's data, WiFi

Master key, as of February 2016, is the world's 5th most downloaded app on iOS and Google Pay store, coming after WhatsApp Messenger, Facebook, Facebook Messenger, and Instagram.

As of June 2016, WiFi Master Key reached 900 million users, second only to WeChat and QQ, becoming the largest mobile internet software tool in China. In the same month, WiFi Master Key also reached 520 million monthly active users, providing over 4 billion daily average connections.

Through cloud encryption technology, WiFi Master Key aims to prevent cyber attacks and provide users with a safe and free connection to the internet. Separately, LinkSure is working with Zhongan Insurance to deploy the WiFi security app for users and hosts to provide additional protection. At the same time, LinkSure is part of the first batch of member companies to enter China Central Government's Internet Safety Union. It is also the only mobile internet company to be part of the Ministry of Industry and Information Technology's Internet safety division in China.

Streamium

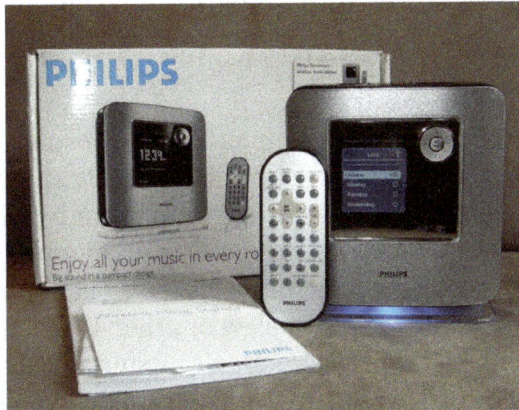
Philips Streamium WAK3300

Streamium was a line of IP-enabled entertainment products by Dutch electronics multi-national Philips Consumer Electronics. Streamium products use Wi-Fi to stream multimedia content from desktop computers or Internet-based services to home entertainment devices. A Streamium device plugged into the local home network will be able to see multimedia files that are in different UPnP-enabled computers, PDAs and other networking devices that run UPnP AV MediaServer software.

Streamium products may also support internet radio, internet photo sharing and movie trailers services directly. Subscriptions to web-based services requiring subscriptions would be managed through the 'Club Philips' portal.

In all cases, using a computer with RSS receiver together with a UPnP AV MediaServer, it is possible to play back audio/video podcast. Some of the popular feeds include

BBC live, Geekbrief, Reuters, Metacafe, YouTube. Although in most cases this video podcaster uses codec formats not supported by Streamium, it's still possible by using software codec transcoders on the PC to convert them to MPEG format.

Philips Media Manager, is—since SimpleCenter version 4— a free open source UPnP AV MediaServer for Windows and Macintosh that is bundled with Streamium. Version 3 of SimpleCenter, was initially developed for inclusion with the Streamium line of products. Since Streamium devices also support photos and videos, SimpleCenter ships with video and image support, under the name 'Philips Media Manager' (PMM).

History

In 2000 Philips' consumer electronics division (business unit Audio) invented the Streamium brand for a "Connected Home". A number of products were released between January 2000 and June 2003. In 2003 the "Connected Home" would be broadened to the "Connected Planet" accompanied by an attempt to steer product development and industrialization from Eindhoven and to include other business units. The "Connected Planet" was less successful, leaving a limited number of products.

Products

The FW-i1000, an audio mini-system including a CD-changer and AM/FM radio, and considered to be the precursor to the Streamium product line, first shipped in June 2001. It had been in development by Philips' audio business group in Sunnyvale (CA, USA) since May 2000. At the January 2001 Consumer Electronics Show they announced and demonstrated the first integrated audio device connecting to "over a thousand internet radio stations". The FW-i1000 was rapidly followed by a slew of other IP-enabled devices. The marketing name "Streamium" and the slogan "Don't dream it, stream-it" was coined and globally registered by Ramon de la Fuente (now at Google) who replaced Tony Cher as a product manager in 2001.

January 2001

- FW-i1000, with iM-networks service offered by Sonicbox, an aggregator of Internet radio service. A limited number of UPnP compliant prototypes where also made for use in the testing of the UPnP AV specifications.

January 2002

- The MCi200, with Philips back-end, several internet based services (and my.philips.com user interface extension protected by a number of patents).

- The iPronto, with WiFi/UPnP, links to security camera and offering a web-based electronic program guide a.k.a. EPG.

January 2003

- The MCi250, an upgrade to the MCi200 with WiFi and UPnP-compliant

Announced at the same event:

- The SL300i and SL400i wireless multi-media adapters

- The MX6000i, with WiFi/UPnP, video, pictures and photo service, music services, 5 DVD changer

- The Streamium-TV, a 32" LCD TV offering similar web-based services as the MX6000i

Both the Streamium-TV and MX6000i were capable of offering video content hosted by a web-based service, the precursor of net TV.

August 2004

- The SL50i wireless PC audio receiver.

September 2004

- The SLA5500 wireless multi-media adapter, to connect to both a PC and an audio system.

September 2005

- The SLA5520 wireless music receiver, the successor of the SLA5500 with access to free Internet radio stations.

June 2006

- WAK700 wireless multiroom music system, with a 40GB hard disk drive to store up to 750 CDs and 3 different ways to listen to music (Listen to different song in each room, take the music from room to room, Simultaneous playback on all stations)

- The WAK3300 wireless music station, with 3 ways to listen to music and 3 different alarm modes.

January 2007

- The WACS7000 wireless multiroom music system, the successor of the WACS700, with a 80GB hard disk drive to store up to 1500 CDs.

June 2007

- The SLM5500 and SLM5520 wireless multimedia adapters, to stream music from PC as the other Streamium devices but also to view pictures and movies stored on your computer on your TV.

January 2008

- The WACS7500 mutliroom music system, the successor of the WACS7500 with color display and access to free Internet radio stations.

April 2008

- The WAS6050 wireless music station, with 4 built-in speakers, and access to free Internet Radio stations.

June 2008

- The NP1100 network music player, to connect to both a PC and an audio system, with wireless access to Internet radio and online music services such as Rhapsody (online music service) or Napster.

August 2008

- The MCI300 wireless micro hifi system

- The MCI500H wireless micro hifi system, with a 160GB hard disk drive to store up to 2000 CDs.

November 2008

- The NP2500 network music player, following the same principles as the NP1100 but with color display, the FullSound technology to restore the details in compressed music, and picture streaming

- The NP2900 network music player, with 4 built-in speakers (no need to connect to an audio system), the LivingSound technology to enjoy immersive music over a wider area and picture streaming.

January 2013

as of January 2013 Philips have abandoned the streamium line without explanation.

It seems that the Fidelio line is to fill the gap streamium left.

Architecture

Whereas the FW-i1000 used the iM-networks service (then known as "SonicBox"), the end-to-end "Streamium" system designed by Daniel Meirsman, included a Philips owned back-end service (the "ECD-interface").

This back-end service allowed Streamium devices to connect to "any number" of web based content delivery services. The back-end would thereby function as a "switchboard" connecting the content delivery services with individual boxes. Moreover, the back-end service would allow Philips to build out an "after-sales" relationship with their customers through the web-based [UI]-extension and would stimulate some early form of web-based social networking with the streamium cafe web site that was set up by Mark Tuttle.

A Philips TriMedia die

A navigation tree would be served to the Streamium-device from the Philips back-end, whereas the content itself would be directly streamed from the service to the Streamium-box subject to the site's policy (subject to the compulsory licences). By manipulating the navigation tree from the front panel of the Streamium-device users could select desired the service, genre, artist, album, track.

From the start, Streamium-devices contained provisions (i.e. an IEEE EUI-64 containing an OUI and a MAC-address, encryption keys, product and software version codes) used to protect streams and support identification mechanisms, as well as allowing downloading of software upgrades (for bug-fixes as well as enabling new features).

In most cases. Streamium functionality was provided by a module (a [PCB]) based on an NXP TriMedia PNX1300. This module implemented:

1. Connectivity to the home network as well as to the Internet (network stack, SAX as opposed to DOM XML-parser, UPnP-stack)

2. Decoding of compressed (audio, image, video) content

A user interface extension was available on the Streamium web-site (my.philips.com) that would allow users to manage their preferences, services and devices.

On the frontpanel or through the on-screen display (OSD) of the Streamium devices, users could mark their favorites or indicate they wanted to learn more about the song being played. The service would then send either an e-mail with more details, or post this info on my.philips.com with a click-through link (i.e. to Amazon.com for purchasing).

Technology Concepts

Under the direction of the Streamium team, Philips CE contributed significantly to both UPnP and Digital Living Network Alliance (DLNA) and other industry efforts.

The Streamiums were extensively used to build advanced concepts, even at other divisions of Philips such as Philips Semiconductors (now NXP Semiconductors), Philips Research and Philips Medical.

Philips Semiconductors would pick up on the vision of Philips Consumer Electronics and task the "advanced system lab" to prototype this vision (first demos end 2004).

Concepts (such as those used in the Connected Home demos) would then be shown by the Streamium team in the "Philips-CE World Tour", an invitation only event at the yearly Consumer Electronics Show in Las Vegas to a selected audience.

- Demonstration of a portable HDD-based and Wi-Fi enabled UPnP renderer (January 2002) for studies of portable entertainment.

- Demonstration of a fully-fledged UPnP enabled network (January 2003) with multiple renderers, servers and control points.

- Bridging between UPnP and Zigbee was demonstrated to control lighting from a UPnP control point as well as a GPRS link that bridged to the UPnP network through a web server (January 2003), linking digital entertainment to home automation and "ambiance control".

- Use cases illustrating the use of RFID in interactive marketing, identification and personalisation (January 2003).

- Tests with the Rhapsody as a UPnP AV-server streaming to an MCi200 (a bit what Sonos is currently offering) (2003).

- Running tests of Apple's Rendezvous protocol on the MCi200 (2003).

- The first demonstration of the NFC smart poster concept with Visa (January 2004).

- Use cases illustrating interoperable DRM (Streamium was one of the devices used to demonstrate the principles underlying the activities of the Coral consortium (January 2004).

- Use cases illustrating the use of recommender technology (January 2004).

- Demonstration (~43 use cases) of a fully connected home including stationary as well as portable devices and mobile phones and executed at the "CES2005 World Tour" by a team of professional actors. (January 2005).

- Use cases in the area of fitness and well-being (January 2005).

- Use cases illustrating uses for UPnP printing (January 2005).

- Experiments with Java-based arcade games.

Services

iM-networks was offered on the FW-i1000 from the start.

The Philips Streamium MX6000i provided traditional home entertainment alongside access to audio-video content from a PC, UPnP mediaserver or online entertainment service. Services included music videos, Web movies, and cinema trailers.

The Philips Streamium service partners were:

- Yahoo LAUNCHcast

- Yahoo! Movies - trailers only

- MP3.com

- Musicmatch Musicmatch Jukebox, Radio MX

- live365.com

- Radio Free Virgin

- Playhouse Radio

- Andante

- Bluebeat

- iFilm - previews and short movies

- Launch - music videos

There was no need to first download files to PC, or even to turn the PC on, to stream multimedia Internet content. (A broadband Internet connection is required.)

A wide array of Yahoo! services including on-demand music videos, movie trailers and clips and photo services was launched together with the Streamium range extension in Spring 2004.

Out of the box consumers would have access to free services. Additionally, a number of services would offer premium (subscription-based) online services allowing consumers to broaden and personalize their home entertainment experience.

Results

In his press conference during the CES2001, Guy Demuynck, then CEO of Philips CE, expressed his vision that the Internet would become as ubiquitous and accessible as the electric grid to devices other than the PC, and that people would increasingly rely on the Internet for information and entertainment.

"The Internet has transformed the way we do business," Demuynck stated, referring not only to communications and promotions but to product design as well. "We intend to put Internet capabilities into many products, making Internet content as accessible as pressing a single button on a TV remote. The Internet dominates all our thinking—it will expand from a browsing activity to an always-on, integral part of daily life," he said.

Innovations

The early Internet audio and Streamium devices had both constant broadband Internet connectivity and a back-end service provided by Philips to aggregate services for its users. In addition, a UI extension was offered (my.philips) that allowed consumers to manage their devices, external service subscriptions, favorites, as well as to add their own streams. The PC-based UI extension was part of the lean-forward/lean-backward approach to this potentially complicated product range:

- Lean backward functionality was accessible on the device directly

- Lean forward functionality was relegated to the PC

In essence, the Philips service offering was a walled garden; but the fact that consumers could add and access their own favorite streams made it more of an open system. Although never published, Streamium relied on an XML-based API—not unlike the APIs we see today published by Web-service providers (YouTube, eBay etc.) -- and implemented XML-based APIs as offered by its service providers.

Commercial Impact

Unfortunately the first Streamium generations lacked support for a number of important media formats such as those included in the Windows Media technologies which hampered their commercial success. Philips' legal team had issues with the non-assertion clause that Microsoft required potential licensees to sign without prior opportunity to check the IP involved. Since Philips did and still does own a substantial patent-portfolio, product management was not allowed to risk signing away rights on a substantial number of important patents.

Consumer Testing

To some extent consumers were involved in some of the design of the Streamiums. Of course, the classical focus test groups were used to find out what features consumers would deem important. But also after the official announcement, a limited number of pre-production versions of the Streamiums would be made available to a limited number of volunteer beta-testers that could sign up via a Philips web-site.

WiFi Explorer

WiFi Explorer is a wireless network scanner tool for OS X that can help users identify channel conflicts, overlapping and network configuration issues that may be affecting the connectivity or performance of IEEE 802.11 wireless networks.

History

WiFi Explorer began as a desktop alternative to WiFi Analyzer, an iPhone app for wireless network scanning that was pulled out from Apple's App Store in March, 2010 due to the use of private frameworks. Since its first release, WiFi Explorer incorporated features that were not included in the last available version of WiFi Analyzer, such as support for 5 GHz networks and 40 MHz channel widths. Starting in version 1.5, WiFi Explorer includes support for 802.11ac networks, as well as 80 and 160 MHz channel widths.

Features

- Displays various network parameters:
  - Network name (SSID) and MAC address (BSSID)
  - Manufacturer
  - AP name for certain Cisco and Aruba devices
  - Beacon interval
  - Mode (802.11a/b/g/n/ac)
  - Band (2.4 GHz ISM and 5 GHz UNII-1, 2, 2 Extended, and 3)
  - Channel width (20, 40, 80, and 160 MHz)
  - Secondary channel offset
  - Security mode (WEP, WPA, WPA2)
  - Support for Wi-Fi Protected Setup (WPS)
  - Supported basic, min and max data rates
  - Advertised 802.11 Information Elements

- Graphical visualization of channel allocation, signal strength or Signal-to-noise ratio (SNR)

- Different sorting and filtering options

- Displays signal strength and noise values as percentage or dBm

- Ability to save and load results for later analysis

- Metrics and network details can be exported to a CSV file format

- Selectable and sortable columns

- Adjustable graph timescales

- Editable column for annotations, comments, etc.

- Customizable network colors

- Full screen mode

- Comprehensive application's help

Limitations

Due to limitations of Apple's CoreWLAN framework, WiFi Explorer is unable to detect hidden networks (except when associated) and does not support external (USB) wireless adapters.

System Requirements

- OS X 10.7 or higher (64-bit)

Logitech Unifying receiver

Logitech Unifying receiver

The Logitech Unifying receiver is a miniaturised dedicated USB wireless receiver which permits up to 6 devices such as mice and keyboards (headphones are not compatible), which must be made by Logitech and of compatible design, to be linked to the same computer using 2.4 GHz band radio communication in a way very similar to, but incompatible with Bluetooth. It is not practical to fit receivers in several computers to allow the same input devices to be used (e.g., with a desktop and a laptop computer), as the devices need to be paired with the receiver each time, although the receiver and input devices can be moved together from one computer to another.

Receivers are supplied with Logitech input devices (e.g., mouse, keyboard, etc.), and are also available separately. The receiver that comes with a Logitech input device is paired with the device at the factory. If you purchase a receiver only (e.g., a replacement), then you are likely going to have to pair it to your existing devices. For that, you will need pairing software, which can be downloaded.

A restriction on some Unifying devices limits them to a maximum of forty-five unique receiver pairings. Once the forty-fifth connection is made, it would no longer be possible to connect such a device to a different receiver. For users who often switch a Unifying device between multiple PCs or laptops with individual receivers, this connection limit will be reached.

Pairing software is available for Windows and Mac OS X from Logitech. Wireless devices using the Unifying Receiver are supported since Linux 3.2. Software to manage Unifying devices on Linux is available from third party developers, for example, Solaar.

BT Smart Hub

The BT Smart Hub (formerly the BT Home Hub) is a family of wireless residential gateway router modems distributed by BT for use with their own products and services as well as wholesale resellers i.e. LLUs. The latest versions of the Home Hub support the Wi-Fi 802.11ac standard, in addition to the 802.11b/g/n standards. All models of the Home Hub prior to the Home Hub 3 support VoIP Internet telephony via BT's Broadband Talk service and are compatible with DECT telephone handsets. Since the Home Hub 4, all models have been dual band (i.e. both 2.4GHz and 5GHz).

The BT Home Hub works with the now defunct BT Fusion service and works with the BT Vision video on demand service. The BT Home Hub 1.0, 1.5 and 2.0 devices connect to the Internet using a standard ADSL connection. The BT Home Hub 3 and 4 models support PPPoA for ADSL and PPPoE for VDSL2, in conjunction with an Openreach-provided VDSL2 modem to support BT's FTTC network (BT Infinity). Version 5 of the Home Hub, released in August 2013, includes a VDSL2 modem for fibre-optic connections. New firmware is pushed out to the Home Hubs automatically by BT.

The Home Hub 5 was followed in mid-2016 by the Smart Hub, a further development of the Home Hub sometimes referred to as "Home Hub 6". It has more WiFi antennas than its predecessor, and a USB 3.0 rather than USB 2.0 port.

History

Prior to release of the Home Hub (2004-2005), BT offered a product based on the 2Wire 1800HG, and manufactured by 2Wire. This was described as the "BT Wireless Hub 1800HG", or in some documentation as the "BT Wireless Home Hub 1800". This provided one USB connection, four ethernet ports and Wi-Fi 802.11b or 802.11g wireless connection. A total of ten devices in any combination of these was supported.

The hardware contained within the BT Home Hub 1.0 and 1.5 was manufactured by Inventel, and is equivalent to other Inventel produced and third-party branded routers such as the Orange Livebox and the Thomson SpeedTouch 7G and ST790. Consequently, the Home Hub 1.0 can be flashed with some firmware such as that for the 7G; however full functionality cannot be achieved using this method. The Home Hub 1.5 firmware, whilst not hardware locked as previously claimed, does have extra locks in the bootloader which can now be circumvented and full functionality achieved.

There are two versions of the BT Home Hub 2.0, the A and the B model. The hardware contained within the Home Hub 2.0A was manufactured by Thomson SpeedTouch, who had bought out Inventel and all their hardware and software rights. This model is electronically identical to the Thomson SpeedTouch TG797n.

The hardware contained within the BT Home Hub 2.0B was manufactured by Siemens' Gigaset division in Germany. The middleware was developed by Jungo, a subsidiary of NDS, and is based on their openRG product.

Some Home Hub 2.0 units were also made by Hon Hai Precision Ind. Co. Ltd.

In the standard firmwares, telnet shell access is available in earlier versions (up to 6.2.2.6) on the BT Home Hub 1.0 with appropriate user permissions. This is identical to the custom shell used in the SpeedTouch range of routers and provides an almost identical software feature set, with a few notable exceptions (e.g. PPP authentication is locked on the BT firmware). This is not the case in the unlocked versions, as full telnet access is available.

There are two versions of the BT Home Hub 3, the A and the B model. The hardware contained within the Home Hub 3A was manufactured by Siemens' Gigaset division (now Sagem) and is based on a Lantiq XWAY ARX168 chipset supporting ADSL2+.

The Home Hub 3B was manufactured by Huawei and also supports ADSL2+. The Home Hub 3B is powered by a highly integrated Broadcom BCM6361 System-on-a-chip (SoC). The BCM6361 has a 400 MHz dual MIPS32 core processor as well as an integrated DSL Analog Front End (AFE) and line driver, gigabit Ethernet switch controller and 802.11 Wi-Fi transceiver.

Features

The BT Home Hub 2.0 was a combined wireless router and phone. It supports the 802.11b/g/n wireless networking standards, and the WEP and WPA security protocols. It supports many of BT's services such as BT Fusion, BT Vision and BT Broadband Anywhere. It can also be used as a VOIP phone through BT Broadband Talk.

The BT Home Hub 3 incorporated WPS functionality, seen on other routers, which enables the user to connect to their encrypted network by the use of a "one touch" button, and also includes "smart wireless technology", which automatically chooses the wireless channel to give the strongest possible wireless signal. WPS has since been (temporarily) disabled by firmware updates due to security issues with the standard.

The BT Home Hub supports port forwarding.

The BT Home Hub versions 3, 4 and 5 may be used for access to files stored on an attached USB stick - USB 2.0 is supported. The server by default has the address File://192.168.1.254 and is available to the entire network.

Hub Phone

The BT Hub Phone is an optional handset that can be bought to work in conjunction with the BT Home Hub 2.0. It calls using the BT Broadband Talk service, and may sit in a dock in the front of the Home Hub or be used on its own stand. It uses Hi-def sound technology when calls between Hub Phones are made. A DECT telephone may be used instead.

With each BT Home Hub released up to 2.0, a new phone model was made to accompany it:

- BT Home Hub 1.0: was supplied with the BT Hub Phone 1010

- BT Home Hub 1.5: was supplied with the BT Hub Phone 1020 (The only difference between the 1010 and the 1020 was the lack of the colour screen and supporting features on the 1020.)

- BT Home Hub 2.0: was supplied with the BT Hub Phone 2.1

- The BT Home Hub 3 and 4 do not work with the BT Broadband Talk service or DECT telephones. After 29 January 2011, BT Broadband Talk was no longer provided as part of BT's broadband packages.

The phones are only partially compatible with newer or older versions of the hub, able to make and receive calls, but with the loss of features including call waiting, call transfer, internal calls, phonebook, call lists and Hi-def sound.

Design

There have been five different versions of the BT Home Hub as of March 2017:

- Version 0.5: grey (no Hub Phone was available, not technically a Home Hub but rather BT Fusion Hub)

- Version 1.0: white (matching Hub Phone was available)

- Version 1.5: white or black (matching Hub Phone was available)

- Version 2.0: black (matching black Hub Phone was available)

- Version 3.0: black (Hub Phones and DECT phones are not compatible) released on 29 January 2011.

- Version 4.0: black (Hub Phones and DECT phones are not compatible) released on 10 May 2013.

- Version 5.0: black, released in mid-October 2013

There were two different versions of the BT Home Hub 2.0: v2.0A (2.0 Type A), manufactured by Thomson, and v2.0B (2.0 Type B), manufactured by Gigaset Communications (now Sagem Communications, Sagem having acquired Gigaset's broadband business in July 2009). Whilst the looks and functionality appear to be identical, the Home Hub 2.0A has been plagued with problems relating to poorly tested firmware upgrades which, amongst other problems, cause the Home Hub 2.0A to restart when uploading files using the wireless connection.

There are also two versions of the BT Home Hub 3: v3A (by Gigaset, now Sagem) and v3B, (Huawei).

The BT Home Hub can only be used with the BT Total Broadband package without modification; the 1.0, 1.5, 2A, 2B and 3A versions can be unlocked. The BT Home Hub configuration software is compatible with both Macintosh and Windows operating systems, although use of this is optional and computers without the BT software will still be able to connect to the Hub and browse the Internet normally.

The 4th generation of the BT Home Hub was released on 10 May 2013. It has been built with a smart dual band technology, making it unique amongst other UK-based ISP provided routers. The Home Hub 4 was supplied free of charge to new customers, with a £35 charge to existing customers. It has intelligent power management technology which monitors the hub functions and puts them individually into power-save mode when not in use. There two variants of the Hub 4, Type A and B.

The 5th generation Home Hub was released in mid-October 2013 and is an upgrade to the Home Hub 4, with Gigabit Ethernet connections, 802.11ac Wi-Fi and an integrated VDSL modem. Customers upgrading from ADSL Broadband pay only a delivery charge;

existing Broadband customers pay a £45 upgrade charge. There are two variants of the Hub 5, Type A with Lantiq chipset (ECI), and Type B with Broadcom.

Models and Technical Specifications

A picture of the rear of the circuit board from a BT Home Hub 3.0 Type A

The BT Home Hub package includes:

- Broadband cable (RJ11 to RJ11)

- Ethernet cable (RJ45 to RJ45) (Cat5e)

- Power adapter

- 2 ADSL microfilters

- Phone to RJ11 converter

- User guide and CD

A USB lead was provided with the Home Hub 1 only.

Spec	BT Home Hub 1.0/1.5	BT Home Hub 2.0	BT Home Hub 3	BT Home Hub 4	BT Home Hub 5	BT Smart Hub
Modem	ADSL2+	ADSL2+	ADSL2+ (PPPoE is also supported in firmware for VDSL2)		VDSL2	ADSL2+ and VDSL2
Wi-Fi	802.11 b/g	802.11 b/g/n	802.11 b/g/n (now with "Smart Wireless", explained above)	2.4 GHz: 802.11n dual-stream 2x2 MIMO. Back compatible with 802.11 b/g. 5 GHz: 802.11n dual-stream 2x2 MIMO. Back compatible with 802.11a.	2.4 GHz: 802.11 b/g/n 2x2 MIMO 5 GHz: 802.11 a/n/ac 3x3 MIMO	2.4 GHz: 802.11 b/g/n 3x3 MIMO 5 GHz: 802.11 a/n/ac 4x4 MIMO
Wireless Security	WEP and WPA-PSK/WPA2-PSK/RADIUS		All previous features but now with WPS (temporarily disabled in firmware updates)	2.4 GHz: WPA & WPA2 (default), WPA, WPA2 and WEP 64/40 5 GHz: WPA2	WPA (2.4 GHz only), WPA2, WPS	

Ports	2x Ethernet 10/100 Mbit/s sockets 1x USB 1.1 socket 2x RJ11 (broadband in and phone)	4x 10/100 Mbit/s Ethernet sockets (RJ45) 1x USB (for network drives) 1x Broadband in (RJ11) 1x Telephone socket	3x 10/100 Mbit/s Ethernet sockets (RJ45) 1x 10/100/1000 Mbit/s GigE Ethernet socket (RJ45) 1x USB socket (now enabled for use) 1x BT Infinity in (RJ45) 1x ADSL Broadband in (RJ11)		4x 10/100/1000 Mbit/s GigE Ethernet socket (RJ45) 1x USB socket 1x BT Infinity in (RJ45) 1x VDSL Broadband in (RJ11)	
Dimensions (w x d x h)	19.5 x 3.9 x 22.5 cm	17.5 x 8.8 x 18.2 cm	18.5 x 4 x 11 cm	23.6 x 3.1 x 11.6 cm		
Software	6.2.6.E or 6.2.6.H	8.1.H.U (Type A), 4.7.5.1.83.3.37 (Type B)	4.7.5.1.83.8.94.1.37 (Type A), V100R001C01B-036SP05_L_B (Type B)	4.7.5.1.83.8.130.1.26 (Type A), FW:V0.07.01.0910-BT (Type B)	4.7.5.1.83.8.204.1.11 (Type A), V0.07.03.814 (Type B)	SG4B10002244 (Type A)

Reported Issues

The security of the BT Home Hub has been questioned by GNUCITIZEN. In October 2007 Adrian Pastor warned the security and BT Broadband community regarding critical vulner-abilities he discovered in the Home Hub. The details of such research were released later in November 2007 and demonstrated how to compromise fully (get root privileges on) the BT Home Hub by tricking a BT Home Hub user to visit a web page crafted by the attacker.

Such research garnered a significant amount of media attention and led to Adrian Pas-tor being invited to BBC Radio 4 where he discussed the issue with Dave Hughes, di-rector of BT Wireless Broadband. Mr Hughes argued that GNUCITIZEN's vulnerability research only covered a theoretical attack. Mr Pastor said that, although GNUCITI-ZEN wasn't aware of such vulnerabilities being exploited in the wild, the attack is fully *practical* as demonstrated by the exploit code released at www.gnucitizen.org. Fur-thermore, Mr Pastor argued that the security of the BT Home Hub wasn't adequate to support the newly introduced Wi-Fi sharing FON service.

BT has the capability to detect remotely and silently all devices connected to custom-ers' networks, and asserts and uses the right to do so, saying that "we don't believe that consent is necessary where the testing is necessary to the service that we are providing."

In May 2017, it was reported that many BT Smart Hub customers are suffering prob-lems with the router constantly rebooting and being unable to maintain a reliable in-ternet connection.

ESP8266

The ESP8266 is a low-cost Wi-Fi chip with full TCP/IP stack and MCU (microcon-troller unit) capability produced by Shanghai-based Chinese manufacturer, Espressif Systems.

The chip first came to the attention of western makers in August 2014 with the ESP-01 module, made by a third-party manufacturer, AI-Thinker. This small module allows microcontrollers to connect to a Wi-Fi network and make simple TCP/IP connections using Hayes-style commands. However, at the time there was almost no English-language documentation on the chip and the commands it accepted. The very low price and the fact that there were very few external components on the module which suggested that it could eventually be very inexpensive in volume, attracted many hackers to explore the module, chip, and the software on it, as well as to translate the Chinese documentation.

The ESP8285 is an ESP8266 with 1 MiB of built-in flash, allowing for single-chip devices capable of connecting to Wi-Fi.

The successor to these microcontroller chips is the ESP32.

Features

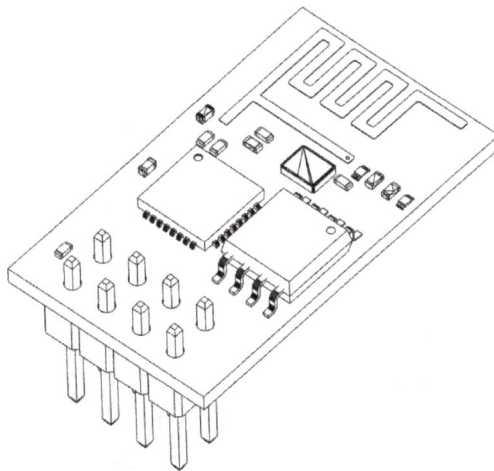

ESP-01 wireframe.

- 32-bit RISC CPU: Tensilica Xtensa L106 running at 80 MHz*

- 64 KiB of instruction RAM, 96 KiB of data RAM

- External QSPI flash: 512 KiB to 4 MiB* (up to 16 MiB is supported)

- IEEE 802.11 b/g/n Wi-Fi

  o   Integrated TR switch, balun, LNA, power amplifier and matching network

  o   WEP or WPA/WPA2 authentication, or open networks

- 16 GPIO pins

- SPI

- I^2C

- I^2S interfaces with DMA (sharing pins with GPIO)

- UART on dedicated pins, plus a transmit-only UART can be enabled on GPIO2

- 10-bit ADC (this is a Successive Approximation ADC)

* Both the CPU and flash clock speeds can be doubled by overclocking on some devices. CPU can be run at 160 MHz and flash can be sped up from 40 MHz to 80 MHz. Success varies chip to chip.

SDKs

In late October 2014, Espressif released a software development kit (SDK) that allowed the chip to be programmed, removing the need for a separate microcontroller. Since then, there have been many official SDK releases from Espressif; Espressif maintains two versions of the SDK — one that is based on FreeRTOS and the other based on callbacks.

An alternative to Espressif's official SDK is the open source ESP-Open-SDK that is based on the GCC toolchain. ESP8266 uses the Cadence Tensilica L106 microcontroller and the GCC toolchain is open-sourced and maintained by Max Filippov. Another alternative is the "Unofficial Development Kit" by Mikhail Grigorev.

Open source SDKs include:

- NodeMCU — A Lua-based firmware.

- Arduino — A C++ based firmware. This core enables the ESP8266 CPU and its Wi-Fi components to be programmed like any other Arduino device. The ESP8266 Arduino Core is available through GitHub.

- MicroPython — A port of MicroPython (an implementation of Python for embedded devices) to the ESP8266 platform.

- ESP8266 BASIC — An open source basic interpreter specifically tailored for the internet of things. Self hosting browser based development environment.

- Zbasic for ESP8266 — A subset of Microsoft's widely used Visual Basic 6 which has been adapted as a control language for the ZX microcontroller family and the ESP8266.

- Espruino — An actively maintained JavaScript SDK and firmware, closely emulating Node.js. Supports a few MCUs, including the ESP8266.

- Mongoose Firmware — An open source firmware with complimentary cloud service.

- ESP-Open-SDK — Free and open (as much as possible) integrated SDK for ESP8266/ESP8285 chips.

- ESP-Open-RTOS — Open source FreeRTOS-based ESP8266 software framework.

Espressif Modules

This is the series of ESP8266-based modules made by Espressif.

Name	Active pins	Pitch	Form factor	LEDs	Antenna	Shielded?	Dimensions (mm)	Notes
ESP-WROOM-02	18	0.1 in	2×9 DIL	No	PCB trace	Yes	18 × 20	FCC ID 2AC7Z-ESP-WROOM02

In the table above (and the two tables which follow), "Active pins" include the GPIO and ADC pins with which you can attach external devices to the ESP8266 MCU. The "Pitch" is the space between pins on the ESP8266 module, which is important to know if you are going to breadboard the device. The "Form factor" also describes the module packaging as "2 × 9 DIL", meaning two rows of 9 pins arranged "Dual In Line", like the pins of DIP ICs. Many ESP-xx modules include a small on-board LED which can be programmed to blink and thereby indicate activity. There are several antenna options for ESP-xx boards including a trace antenna, an on-board ceramic antenna, and an external connector which allows you to attach an external Wi-Fi antenna. Since Wi-Fi communications generates a lot of RFI (Radio Frequency Interference), governmental bodies like the FCC like shielded electronics to minimize interference with other devices. Some of the ESP-xx modules come housed within a metal box with an FCC seal of approval stamped on it. First and second world markets will likely demand FCC approval and shielded Wi-Fi devices.

AI-thinker Modules

AI-Thinker ESP8266 modules (ESP-12F, black color) soldered to breakout boards (white color)

These are the first series of modules made with the ESP8266 by the third-party man-ufacturer *AI-Thinker* and remain the most widely available. They are collectively re-ferred to as "ESP-xx modules". To form a workable development system they require additional components, especially a serial TTL-to-USB adapter (sometimes called a USB-to-UART bridge) and an external 3.3 volt power supply. Novice ESP8266 devel-opers are encouraged to consider larger ESP8266 Wi-Fi development boards like the NodeMCU which includes the USB-to-UART bridge and a Micro-USB connector cou-pled with a 3.3 volt power regulator already built into the board. When project develop-ment is complete, these components are not needed anymore and it can be considered using these cheaper ESP-xx modules as a lower power, smaller footprint option for production runs.

Other Boards

The popularity of many of these "other boards" over the earlier ESP-xx modules is the inclusion of an on-board USB-to-UART bridge (like the Silicon Labs' CP2102 or the WCH CH340G) and a Micro-USB connector coupled with a 3.3 volt regulator to provide both power to the board and connectivity to the host (software development) computer commonly referred to as the console. With earlier ESP-xx modules, these two items (the USB-to-Serial adaptor and a 3.3 volt regulator) had to be purchased separately and be wired into the ESP-xx circuit. Modern ESP8266 boards like the No-deMCU boards are a lot less painful and offer more GPIO pins to play with. Most of these "other boards" are based on the ESP-12E module, but new modules are being introduced seemingly every few months.

Nexus Hawk

The Nexus Hawk 4G is a gateway router linking broadband cellular data, such as CDMA, GSM and Wi-Fi (IEEE 802.11)a, b, g, n) and WAN (such as BGAN Satellite) networks providing enterprises with broadband wireless internet/network data services in mo-bile and remote environments.

The Nexus Hawk's original development was funded under a DOD prime contract. The technology was primarily designed for military use and supports public safety. The Nexus Hawk is currently in use by law enforcement agencies, governmental data infra-structure, commercial fleet, connectivity in and to retail locations, and livery services in Washington DC.

The device provides; secure access to public and private wired and wireless networks including; Sprint (CDMA EVDO Rev A, 1xRTT), Verizon Wireless LTE, CDMA EVDO Rev A 1xRTT, AT&T Wireless 4G, GSM /HSDPA, Telus HSDPA+, CDMA EVDO Rev A 1xRTT, Washington DC EVDO Rev A Regional Wireless Broadband Network (RWBN), non-U.S. cellular networks, and secure WiFi. GPS for applications such as Automatic Vehicle Location (AVL) sometimes commercial referred to as fleet tracking or Geo-based Dispatch and Navigation. Connectivity to multiple simultaneous WAN via GIG ethernet, USB or WiFi paths with user-selectable order for failover and fail back. Access to 4 simultaneous WANS and GPS. Automatic and persistent network connections. Incorporates 2 USB and 4 PCI-M slots to accommodate future networks such as WiMAX and Public Safety Band), accepts ExpressCard 34mm air cards, PCMCIA CardBus air cards and USB air cards via adapter, Secure Remote Configuration Management, Built in IPsec and OpenVPN and pass through security features, FIPS140-2 SSL Certified Module.

Prism (Chipset)

The Prism brand is used for wireless networking integrated circuit (commonly called "chips") technology from Conexant for wireless LANs. They were formerly produced by Intersil Corporation.

Legacy 802.11b Products (Prism 2/2.5/3)

The open-source HostAP driver supports the IEEE 802.11b Prism 2/2.5/3 family of chips.

Wireless adaptors which use the Prism chipset are known for compatibility, and are preferred for specialist applications such as packet capture.

No win64 drivers are known to exist.

Intersil Firmware

- WEP

- WPA (TKIP), after update

- WPA2 (CCMP), after update

Lucent/Agere

- WEP

- WPA (TKIP in hardware)

802.11b/g Products (Prism54, ISL38xx)

The chipset has undergone a major redesign for 802.11g compatibility and cost reduction, and newer "Prism54" chipsets are not compatible with their predecessors.

Intersil initially provided a Linux driver for the first Prism54 chips which implemented a large part of the 802.11 stack in the firmware. However, further cost reductions caused a new, lighter firmware to be designed and the amount of on-chip memory to shrink, making it impossible to run the older version of the firmware on the latest chips. In the meantime, the PRISM business was sold to Conexant, which never published information about the newer firmware API that would enable a Linux driver to be written.

However, a reverse engineering effort eventually made it possible to use the new Prism54 chipsets under the Linux and BSD operating systems.

References

- Andy Ihnatko (June 8, 2010). "Apple keynote fail. Google keynote fail. It's all good.". Chicago Sun-Times. Archived from the original on June 11, 2010. Retrieved June 9, 2010

- "Five Hundred Wi-Fi Networks Walk into a Bar | the Blog | Future Tense with John Moe | American Public Media". Futuretense.publicradio.org. Retrieved 2010-06-08

- Graves, Brad (November 4, 2016). "Novatel Wireless Changing Name to Inseego Corp.". San Diego Business Journal. Retrieved 7 November 2016

- Brian Nadel (November 4, 2011). "Wi-Fi tethering 101: Use a smartphone as a mobile hotspot". Computerworld. Retrieved 2012-01-16

- Danny Briere; Pat Hurley; Edward Ferris (2008). Wireless Home Networking for Dummies (3 ed.). For Dummies. p. 265. ISBN 978-0-470-25889-7

- Kim, Eugene and Alex Colon, "The Best Mobile Hotspots of 2015", June 10, 2015, PC Magazine retrieved November 4, 2015

- Graves, Brad (September 22, 2016). "Novatel Sells Wireless Hotspot Business for $50 Million". San Diego Business Journal. Retrieved 23 September 2016

- Nedevschi, Sergiu (2008). Maximizing Performance in Long Distance Wireless Networks for Developing Regions. University of California, Berkeley: ProQuest. p. 28. ISBN 1109096100

- "$10 wok keeps TV station on air". The New Zealand Herald. 2007-02-22. Archived from the original on 2012-09-05. Retrieved 2008-03-18

Permissions

Index

www.ingramcontent.com/pod-product-compliance
Lightning Source LLC
Chambersburg PA
CBHW062005190326

41458CB00009B/2970